THE COLORS OF GENES
or the perversion of genetics

TRANSLATED FROM FRENCH
BY THE AUTHOR

© 2020 by Majambu Mbikay

BIOMESC
Consulting Incorporated

MAJAMBU MBIKAY
Pharm., M.A., Ph.D.

THE COLORS OF GENES
or the perversion of genetics

Facts, fantasies, and fallacies
about the alleged inferior intelligence
of Afro-descendants

BIOMESC

By the same author

- *Je saurais croire. Réflexions sur la science, la foi et la société.* Assay, 74 pages. Muhoka Editions, 2005.
- *La voie de l'exil. Murmures et confidences d'un Africain Canadien.* Epistolary novel, 290 pages. Lulu Press, 2009.
- *Proprotein Convertases* (coedited with Nabil G. Seidah). A collective book about converting enzymes, 367 pages. Humana Press, 2011.
- *Demain, le Congo. La République démocratique du Congo est-elle un artefact ?* A sociobiology assay, 113 pages. Lulu Press, 2012.
- *As I Know, So Shall I Believe: An African Scientist's Musing on Beliefs, Science and Society.* Assays. 79 pages. Lulu Press, 2012.
- *Repères.* Assays, 109 pages. Akulà Initiatives, 2013.
- *À tout propos, et des mots pour le dire.* Assays, 235 pages. Akulà Initiatives, 2014.
- *Entre le rêve et le souvenir.* A romanced autobiography, 317 pages. Akulà Initiatives, 2014.
- *Concours & Circonstances, Tome 1, Naître et grandir au Congo.* 247 pages. Akulà Initiatives, 2015.
- *Signposts of Hopes and Illusions.* Assay, 104 pages, Akulà Initiatives, 2016.
- *Concours & Circonstances, Tome 2*, Contributions et héritages. 534 pages. Akulà Initiatives, 2018.
- *La couleur des gènes: la perversion de la génétique.* BIOMESC Consulting Inc. 2018.

*Kamenga, my son,
know that there is more sublime to you
than the simple sum of your genes.*

MAJAMBU MBIKAY, Pharm. D. (Pharmacy, Lovanium University, 1971), M.A. (Biochemistry, State University of New York at Buffalo, 1975), Ph.D. (Biochemistry, State University of New York at Buffalo, 1979), is a retired Professor of Biochemistry and Molecular Biology at the University of Ottawa and a Biomedical Researcher at the Ottawa Hospital Research Institute as well as the Clinical Research Institute of Montreal. The central theme of his long scientific career has been the way Nature increases its capacities by creating a variety of molecules through fragmentation of proteins, at the last stage in the flow of genetic information. His more recent research has focused on the genetics of susceptibility and resistance to chronic and infectious diseases.

Professor Mbikay was born and raised in the Democratic Republic of Congo.

TABLE OF CONTENTS

FOREWORD 11
FOREWORD TO THE ENGLISH VERSION 21
INTRODUCTION 23
CHAPTER 1 CONCEPTS OF HEREDITY IN AFRICAN TRADITIONS 29
    1.1. Heredity observed 29
    1.2. Heredity explained 33
    1.3. Morbid heredity 36
    1.4. Unusual births 39
CHAPTER 2 FROM HEREDITY TO GENETICS 43
    2.1. The monk's garden and the great theory 44
    2.2. Blueprints to bequeath 46
    2.3. The bricks of the genome 53
    2.4. The motions of the genome 58
    2.5. The fingerprints of environment 60
CHAPTER 3 THE BRIEF HISTORY OF THE HUMAN GENOME 63
    3.1. Human life is a wink 63
    3.2. These Africans that we all are 66
    3.3. Abebe and Abeba: the seminal couple 70
    3.4. The signatures of natural selection 73
    3.5. The illusion of race 78
CHAPTER 4 THE PERVERSION OF GENETICS 83
    4.1. Eugenics gone amuck 83
    4.2. Obsessing about IQ 88

4.3. But is it true?	95
4.4. The assertions of genetics	107
4.5. The silence of genes	118
CHAPTER 5 LOOKING TO THE FUTURE	125
5.1. Genetics must not be a religion	125
5.2. Africans beyond genetics	131
NOTES AND REFERENCES	139
LEXICON OF RECURRENT GENETIC TERMS	147
INDEX	151

Foreword

SEVERAL INCIDENTS in my professional career have led me to write this book. The North American reality has made it that, in my field of interest, the biology of genes or genetics, researchers of African ancestry have, until very recently, been an exception. In almost every major international conference I attended during my scientific career, I was a curiosity. The presence of other visible minorities, of Chinese or Hindus, was considered normal. Not mine. I was an anomaly. Many colleagues, including eminent scientists, have reminded me of this without malice on multiple occasions. I will describe two incidents as examples.

In November 1992, in Denver, Colorado, I attended a mega-congress gathering nearly 6000 scientists from all over the world. Luc Paquet, one of my doctoral students, accompanied me there. He noted that nearly all the African-Americans we met in the hallways of showrooms and who worked there, most of them as waiters and sweepers, greeted me with a furtive and whispered *'Hi Brother'*. He wanted to know why. "They feel relieved and comforted to see one of their own in this gathering of scientists," I replied, smilingly. On my return to Montreal, I told the anecdote to the late Bijimine Mputu Grégoire (1945-2015), the first lawyer of Congolese origin in Quebec and a close friend. With his typical verbal verve, Mputu explained to me that this greeting stemmed from what he called a *persuasion effect:* "If you were able to end up in this place, it means that we or our descendants can too." According to Mputu, this is what my fellow African-Americans meant by this *'Hi Brother'*. I was a breach in the wall of doubt, anxiety, and fear, fear of being frozen by biology into a status of second-class citizens, if not of sub-humans.

Three years later, March 1995, in Tahoe, California, at a Keystone Symposium attended by a small number (approximately 200) of molecular biologists, I had the honor of exchanging for a few minutes with none other than Dr. Walter Gilbert of Harvard University, a winner of the 1980 Nobel Prize in Chemistry, which he had received for developing a chemical method of determining the succession of units making up DNA. During the first morning session of the first day, we sat next to each other. I had noticed that he looked at me intermittently, but piercingly. At the 10 o'clock coffee break, sneaking between attendees, he resolutely approached me, read aloud (and well!) my name on my identity badge, and, without detours, asked me wherefrom I came. When I showed him the badge that identified the *Institut de recherches cliniques de Montréal* (IRCM) as my home institution, he insisted that I tell him my nationality of birth. I informed him that I was born, raised, and educated in Congo-Zaire. "I knew it! You aren't an African-American," he said in a satisfactory tone. The conversation moved on to science, but before he left me, he told me of his profound joy at finally seeing a person like me ('*one of you people*' were his exact words) in this kind of forum. Dr. Gilbert may have forgotten the remark; not I, obviously. That morning, I may have ever so lightly answered his lingering questions about the absence of 'us people' in the sciences; I may have consoled him out of his unspoken sorrow about the predicament of people of African descent.

The two events described above illustrate the questions of non-Africans about Africans. The next two deal with what I dare call '*Africans' fear of genetic discourse*'. They occurred more than 20 years apart in less formal contexts of my interactions with a friend in one case, and with one of my children in the other.

Foreword

In September 1983, having completed my postdoctoral training at the University of Sherbrooke, I accepted the position of Senior Scientist at the IRCM with an affiliation as a Research Assistant Professor at the University of Montreal. Shortly after, I met a colleague whom, for not having asked for his consent in the narration of his life story, I identify here by the initials KV. In his early 40s, Canadian of Nigerian origin, KV had lived two thirds of his life outside Nigeria, in England for twenty years, and the rest in Canada. Economical of height and built, he nevertheless exuded an aristocratic aura that was reinforced by the Oxford's accent which modulated his verb. KV was a biologist by training, but his Ph.D. degree obtained from the prestigious McGill University had only temporarily afforded him a scientific career worthy of the title. While searching for better opportunities, he had fallen back on social work to the benefit of Montreal's Anglophone Afro-Caribbean youth for a modest and occasional stipend. Fortunately, he was married to a lady from the Quebec majority, whose professional career could support him and their two children, a boy and a girl. Along with a few West Indians and Africans, KV had set up a *Canadian Association of Third World Scientists* (CATWS). It was as members of this association that we met and became friends. Very quickly, he recruited me into his social initiatives and entrusted me with the Francophone branch, which, until then, was less active than its Anglophone counterpart. The work consisted of tutoring, career counseling, health education, and more. I also chose to serve as a scientific advisor to the *Canadian Sickle Cell Anemia Society*. Sickle cell anemia (SCA) is a genetic blood disease that predominantly affects people of African ancestry, a minority group in Canada. Little known in the health care system of the time, SCA was subject to the most extravagant explanations. My role was to educate parents about the genetic basis of this disease and current efforts to find a therapy. In this context, I had occasionally given interviews on the subject on community radio and television programs. Very quickly,

KV and I discovered how difficult it was to get parents of SCA children to talk about their experience in the public forums we organized. They were willing to do it privately, but not publicly. Obviously, *they considered the disease a shame, a stigma, and a curse that befell Africans and their descendants*, a premise and prejudice fatally internalized.

In 1990, KV lost his father in Nigeria. He went there to pay tribute to his memory and manage his succession. For the father, in addition to being a great pastor of the Adventist Church, was also a wealthy landowner. Back in Montreal, KV shared with me the joys, surprises and disappointments of his stay in his native country. He told me how the State and the people had tacitly agreed *not to disturb each other* in the national swamp in which they waded together; how, during a conversation with Nigerian intellectuals about the generalized slump in which sub-Saharan Africa seemed engulfed, he was asked, as a biologist, *whether melanin was a neurotoxin!* On this question, KV stopped speaking, too indignant to continue. To trivialize the matter, I pointed out to him, in an amused tone, that tyrosine, one of the units of protein chains, was also the raw material for the manufacture of both melanin and a chemical messenger of the brain (a neurotransmitter) called dopamine; that, perhaps, an overproduction of melanin was accompanied by a brain deficit in dopamine. I was joking, of course; but the dark stare on KV's face quickly made me realize that the subject was no cause for laughter; that I was carelessly playing the game of those who distort science to consolidate prejudices and to justify societal injustices. The joke was a way of soothing my pain that such a question ever came to be asked.

This brings me to the second determining incident in my decision to write this book. It revolved around the controversy raised by Dr. Pierre Mailloux during his appearance on the Radio Canada television program *"Tout le monde en parle"* (Everybody is talking about it) on September 25, 2005. This psychiatrist, with his abundant beard of Coptic pontiff, had made a name for himself on radio and television as a host and popularizer on mental health issues. He was reputedly not one to bite his tongue. His outspokenness, deliberately outrageous, had brought him praise and criticism in the past, but nothing compared to the outcry that his September 25, 2005 comments would raise when he said that *scientific studies had shown that Africans and Amerindians had a lower IQ than Caucasians*. The outcry came mainly from ethnic interest groups such as the Black League and the Société Saint-Jean-Baptiste. These organizations were attacking, not only the doctor but also the Radio Canada for having given him such a public forum for his inflammatory comments. They accused the radio station of wanting to recklessly raise its ratings on the pretext of respecting freedom of speech.

For my part, Mailloux's words left me superbly indifferent. I didn't find anything new there. Heard it, seen it, in many forms and circumstances. Mailloux was not the first to think it out loud; he wouldn't be the last either. I even found inappropriate the political correctness, which would have confined to silence controversial opinions and convictions that secretly comfort our various psychological complexes. "What's bred in the bone comes out in the flesh", it is said. Transparency of speech on this subject was much healthier than any hypocritical 'mum's the word'.

"Let Mailloux shout himself hoarse in empty verbiage for his cheap pleasure; let him nourish his petty illusions of importance. Besides, what does it matter? Tomorrow, the sun will rise as it did yesterday. Tomorrow, as yesterday, some humans will love one another, others will kill each other. High IQ or low IQ, with machetes or Kalashnikov, in the hut or skyscraper, by pirogue or yacht, we are but animals in the grip of our instincts, trying as best we can to rise above them, in our constant quest for the humanity in us." That was my attitude. Cynical, wouldn't you say?

This indifference was broken by an emotional appeal by my son Kamenga, whom the show seemed to have deeply upset. He came to see me insisting that, as a scientist of African descent, I ask for a right of reply in the same program to contradict Mailloux's comments and destroy his arguments. Kamenga's insistence was out of character for this son of few words, quiet and reserved (except on stage when he explodes in his rap poetry). I tried to interpret his emotion. Of the five children I have raised in Quebec, Kamenga is the most immersed in Quebec's culture. He literally lives it and in it; he knows its manners and customs, its language and jargon. He even started a family with Nancy Brière, a native of Quebec. I inferred that the allegations of Pierre Mailloux, a medical authority, on the intellectual inferiority of Africans and Afro-descendants, had hurt him in particular because they disparaged him in the eyes of his Caucasian wife and friends. He needed a counter-authority to restore the *'truth'* and safeguard his *'dignity'*. Who else to do it right than his own father?

I didn't ask Radio-Canada for the right of reply, but I invited Kamenga to an introspection of his life experience. I assured him that he wouldn't find in it any evidence of any inferiority to anyone, the human soul being unable to concede to the logic of such a comparison. I explained to him that, like every human being, he had distinctive character traits that made him neither more nor less pertinent as a human being; that, like all human beings, he has experienced successes and failures; that, surely, he knew of many others, all origins combined, who were, in many respects, less endowed and less fortunate than him.

Kamenga listened to me, annoyed and resigned. Apparently, he didn't know what to do with my philosophical exhortations. He would have preferred that I throw myself into the public debate. This option, unfortunately for him, would have been uncharacteristic of me. Because of the emotions that usually accompany them, public debates have always appeared to me as brutal battles of slogans. They reinforce prejudices rather than change opinions. I argue better with a rested head, preferably on paper. I carefully read the books of Mailloux's ideological compadres. It is thus on paper that I now intend to answer the insistent questions that my son put to me on that day.

I chose the title and the first lines of this book in the waiting room at the Maisonneuve Hospital of Montreal, while Andy Kadima and Keisha Mbiya, the son and daughter of Kamenga and Nancy, were having a preventive tonsillectomy. I showed Kamenga three possible titles for the book: '*Black Genes*', '*Racist Genetics*', or '*The Color of Genes*'. The last title was his favorite; mine too. It was intriguing and attractive enough without being provocative or outrageous. It appealed to common sense by its absurdity and could attract, without undue prejudice, the curiosity of both parties in the antagonistic discourses on the subject. Perhaps it would persuade them all to peruse the pages of the book and learn from them. I have consented to this title for a practical reason too, I who have resolved to never again make concessions to a nonsensical semantics, based on names and adjectives loaded with subliminal messages.

The role of '*slayer of scientific myths*' that I assume in this book has been imposed on me by the circumstances of my career. I did not actively pursue it. By accepting it, I have burdened myself with anxieties which I would have gladly done without. I would have preferred to be completely ignored as an African and be recognized merely as a scientist who loves his job and tries to enjoy it as much as he can. I would have liked to write a book on the meanderings of genes and their sorting by natural selection. But as years went by, besieged by the looks and questions of my fellow Africans, I slowly began to gather the bibliography focused on the historical, sociological and scientific foundations of *genetic mythologies*, while reflecting on the best ways to expose and unmask them for the largest public.

Indeed, this book has no pretension of erudition; it is an essay and not a treatise: it attempts to gather existing data and incorporate them into a coherent and comprehensible vision for the reader unfamiliar with the parlance of the *'temple'*. This approach requires multiple stops to avoid the pitfalls of oversimplification and to satisfy the need for clarity, making the best use of this imperfect tool we call language. Well aware of the fact that, in the minds that welcome them, the most precise words all too often take on the meaning that prejudices and emotions impress upon them, my deepest wish is that what I write in this book could be understood as I meant it.

Obviously, I set myself up for an ambitious challenge. Only the reader will tell you how well I was able to meet it. If that is to be, it will also be partly thanks to the careful reading of my benevolent reviewers. I thank Professor Nkongolo Kabwe Constant of Laurentian University in Sudbury, Ontario, a fellow geneticist, for his 'peer review' of the draft of this manuscript. His expert comments allowed me to further clarify some biological concepts unfamiliar to the layperson.

In the course of writing this book, I often solicited the medical opinions of my three professional cousins, dear accomplices since our university years in Kinshasa: Drs. Likongo Yona, Tshibemba François, and Ngoy Fabien. Together, they total up almost 100 years of medical experience in Africa, where they have seen and heard the fears of heredity on different faces and in different languages. Herein, I have borrowed and translated their thoughts on the promises and risks of genes for health.

Again and again, I would like to thank the generous reviewers of all my books, my longtime friends, all professional educators, Drs. Biakabutuka Rémi, Kapanga Kapele Charles and Ndia-Bintu Kayembe. They have enriched my discussion of intellectual quotients with their comments, and have collectively contributed to the literary quality of this work. Their diligence in this project reinforces our old friendship, but doesn't release me from my duty of gratitude to them.

Finally, my warmest thanks go to Mujangi-Annie Lunganga, my wife, who deigned to carry out with me the most beautiful human genetic experiment of all, as she chose to combine her chromosomes with mine and, for more than 49 years, has consented to admire by my side the hatching and blooming of the five fruits of this sublime apposition, each as unique as it was original: our offspring.

FOREWORD
TO THE ENGLISH VERSION

SINCE ITS PUBLICATION in February 2018, *"LA COULEUR DES GÈNES"*, the original French version of this book, arose interest beyond expectations from people in all walks of life, in the Americas, Europe, and Africa. Many insisted that I write an English version of it to reach even more readers. I conceded to the request, well aware of the fact that translation is a dialogue between cultures; that it implicitly involves some degree of betrayal of the original text, of its idioms, sensibilities, and cultural innuendos. This is to say that, although I have attempted to remain as faithful as possible to the 'spirit' of the French script, I often found it necessary to revamp the phraseology in conformity with what I judged to be an acceptable English style, to squarely substitute idiomatic expressions of Molière's language with those of Shakespeare's, humbly recognizing that I would never approach the stylistic mastery of either of these two illustrious names.

My life path has made me a somewhat 'multicultural man', navigating with relative ease between Congolese, French, and English cultures. In my writing, I often can't resist the temptation or deny myself the pleasure of mixing words and phrases from my alternate cultures when I 'sense' that they express or synthesize my thoughts better. I realize that this 'game of words' carries on occasions the risk of creating 'intercultural interferences' as well as textual and contextual misinterpretations in the mind of the reader without the appropriate cultural backgrounds. On such occasions, I have used liberal translations, periphrases, or paraphrases to clarify my thoughts. In the same vein, quite reluctantly, I removed the African proverbs that concluded each chapter that some readers found distracting.

I would like to thank all the readers of the French version of the book for their comments, questions, and suggestions. To adequately address them, I have made a few additions to the present English version. More specifically, I have introduced the concept of alternate splicing as a source of biological diversity (Chapter 2.3); I have inserted recent claims of an impact of the Neanderthal heritage on brain structure and cognition (Chap 4.3); I have illustrated the segregation of genetic diseases among populations with a few more examples (Chapter 4.4) and further clarified the far-reaching implications of the CRISPR/Cas9 technology for the future of the human genome (Chap. 4.4); I have converted the Conclusion into Chapter 5 and given it a title (Looking to the Future). Therein, I explicitly defend my views on cultural adaptation as a gateway to development.

Although I defined genetic terms at their first use, some readers of the French version found it hard to recall their meanings as they went on through the pages, necessitating frequent returns to earlier ones. To spare them such an inconvenience, I have placed after the concluding chapter a short lexicon of recurrent genetic terms listed in alphabetic order.

I passed this manuscript by the sharp eye of James Rochemont, a.k.a. Jimmy, my good friend and colleague for so many years, whose command of the English language, of the proper word and right turn of phrase, I have always admired. His 'imprimatur' on the content and style of this book was a reassuring compliment.

Lastly. I am grateful to my eldest son, Tshibangu K. Alex, an avid dilettante reader of everything scientific, for his many inquiries on the content of this book. His questions have allowed me to answer in anticipation those of my future readers.

INTRODUCTION

THIS BOOK IS A TENTATIVE ANSWER to the numerous questions that, because of my profession as a molecular biologist and geneticist, I have faced on the part of my fellow sub-Saharan Africans, the so-called 'Black' Africans. These questions reflect a persistent concern about the term 'genetics', as if the word carries an intimidating fatality, a kind of terrible secret that it would be preferable not to reveal for fear of explaining and justifying the perceived precariousness of African peoples' condition.

The concern stems from the *contemporary mythology of science as the ultimate reference and authority for objective truth*. This mythology permeates the public discourse and scenery. It is invoked as the final argument to truth. The word DNA has become the key to the ultra-secret vault where is kept the secret of every human being: of his origins and future, of his station and becoming. Scientists, who have the privilege of access to this "Holy of Holies", come out of it with revelations that define the destiny of every human being: his birth, growth, health, disease risks, longevity. Genes have become the oracles of modern times: they are consulted, listened to, and blindly believed; their dictates are enthusiastically endorsed or fatalistically endured; they are invoked to explain and justify all and everything biological.

This work is an effort to deconstruct the myth, an attempt to place genetics in a broader context of interpretations of the biological destiny. Genetics is something, but it is not everything. It explains some things, but its explanations remain partial and often provisional. But the most important point of my dissertation is that, by its nature and in its mechanisms, genetics abhors fixity and rigidity. It is therefore completely absurd and fundamentally fallacious to attempt to derive sociological assumptions from it, to rank individuals or groups of individuals by their differences.

In this book on genetics, I will obstinately refrain from using categorizations of individuals based on color; firstly, because they are imprecise and foolish; secondly, because they translate semantic laziness, a submission to outdated linguistic conventions, grossly disfigured by history. I find it more appropriate to refer to human groups by their geographical ancestral origins, as established by prehistorical and historical migrations which, as we shall see, have left their marks on genes. Henceforth, those commonly referred to as *Negros or 'Blacks' and their descendants, will be called Africans; those dubbed 'Whites' and their descendants will be called Europeans or Caucasians; and those broadly labeled Mongoloids or Orientals and their descendants will be simply called Asians.* The adopted nomenclature is semantically imperfect, but it has the merit of being emotionally indifferent.

I have divided this essay into five chapters. In the first chapter, I will analyze the popular concepts of heredity – the mother of genetics – in African cultures. To do so, I will rely primarily on my own Baluba culture from the central region of the DRC in which I was brought up and which I know best. I will try, whenever possible, to enrich this analysis with perspectives drawn from other African traditions, as described in scientific documents that I was able to consult. The African bias is intentional since this work is an attempt to address the many questions on heredity and genetics in the mind of Africans.

In the second chapter, I will describe the science of genetics, as it is understood in the era of DNA. Convinced that any scientist who knows his subject must be able to convey its content to an eight-year-old child, I will present the ins and outs of modern genetics in terms that are accessible to the moderately educated majority while avoiding oversimplification and caricature. Since genetics has a history, I find it indicated for the reader's scientific education to point out conceptual differences and controversies it has experienced in its relentless pursuit of plausibility. As a branch of science, genetics is a product of culture; it is also a producer of culture. It has its rules of rigor, reproducibility, and coherence, true, but these rules do not free it from the influences of the surrounding environment. Sometimes this environment makes it say what it doesn't say and doesn't know. Young still, it awkwardly defends itself and has a hard time making its voice heard. This book lends a voice to its true message.

In the third chapter, I look back on the long history of life from which our short story as *Homo sapiens* sprang. This history is nothing more than the history of genes, genes that 'live' and 'want to endure' against all odds; genes that unite all living things in a *Tree of Life*, weaving, growing, sharing, and enriching each other, awakening, dozing off, or slumbering in the inner and outer environments; and above all, genes that carry countless marks of the meandering paths of life, bear witness to these paths, and tell us about them; genes that recount to us who we are, what we can be, where we come from, and, very possibly, where we are going.

In the fourth chapter, I will expose the myths arising from the advances in genetics, more specifically, myths surrounding the theory of evolution and hereditary differences in intelligence among human groups. These myths have all too often been used, with or without malice, to explain the inequalities between individuals or societies, and to cast the differences into biological fatalities. I will uncompromisingly denounce the tendency of some academics, including a few Nobel Prize winners, to '*scientify*' their opinions, using their aura of renowned scientists to bestow credibility to their prejudices or to speculate too freely on subjects that have not been subjected to the test of scientific verification. For science, by definition, is an exercise, not of opinions, but of facts. I will establish the facts as genetics has revealed them; I will denounce the falsehoods attributed to it.

In the fifth and concluding chapter. I will examine the reasons for this thirst for scientific revelations that grips public opinion in a world that has become skeptical of traditional religious 'truths'. This rush to '*sacralization*' distorts the mission of science. As already mentioned above, science has nothing to do with truths; it has everything to do with rational and verifiable answers (sometimes useful, but often provisional) to the questions and challenges raised by our perception and experience of reality. I will also question the insidious need (which may be driven by atavistic instincts) that man seems to feel to evaluate species, groups, and even individuals, in a sort of '*evolutionary scale of worthiness*', using genetics as an interpretative framework. This more or less narcissistic need is often expressed by members of groups who believe to have been favored by evolution, because of their current social, economic, technological or military supremacy; of their ability to master matter, time and space. The reality of our time is that, among the peoples of the world, peoples of African descent, by their present condition, readily lend themselves to denigration by public opinion, often tacit, but sometimes vocal. I will try to explain this condition, not to dismiss, justify or excuse it. Nothing can better rehabilitate Africans than a renaissance of their continent through an adequate adaptation to the ecology of modern times and by the intelligent exploitation of the multiple survival tools that this ecology offers.

Finally, I will speak briefly about the damage, sometimes deadly damage, caused by misinterpretation and misuse of genetics. In the course of History, humans have taken particularities resulting from genetic assortments for a divine decision, a blessing, or a curse. They have used them as pretexts for arrogance, malice, fear, and even defeatism. Indeed, if there is one thing that genetics teaches us, it is that, in spite of all our differences, we are all, each in his unique way, *'success stories'*, resounding testimonies of the useful and winning expression of genes in environments that have their particularities, opportunities, and constraints.

CHAPTER 1

CONCEPTS OF HEREDITY IN AFRICAN TRADITIONS

1.1. Heredity observed

HEREDITY IS THE TRANSMISSION of the traits of a group of living beings from one generation to the next; in fact, it is the reproduction of the group, in the literal sense of the word; it is the passing on of the traits of this group, from ancestry to progeny. The traits as well as the ability to pass them successfully from generation to generation, define the group. In this process, members of groups make use of mechanisms adapted to their nature and deployed naturally. All the peoples of the world know this. Observing the transmission of physical traits in the plant and animal kingdoms,[1] peoples also recognize that, within these kingdoms, reproductive groups are the ones that shared the higher number of common traits. If plants engender plants, animals engender animals, and fish engender fish, the groups that make up these kingdoms cannot combine and jointly transmit their traits to their progeny unless they belong to the same reproductive group. *This inbreeding capacity among members of a group defines the species.* Thus, among fish, sea bream and sardines cannot share their traits, neither can ducks and chickens among birds, nor snakes and lizards among reptiles, nor leopards and lions among felines.[2] The reason is that they belong to different reproductive groups, to different species. All the peoples of the world empirically know this, by merely observing Nature.

On the other hand, within each species, there are subgroups with distinctive traits recognizable at first sight. These are called *'types'*. Indeed, geographical or sociological isolation can lead to the fixing of some unique traits. Inversely, geographical or sociological proximity could allow for trait sharing and mixing among subgroups, since they belong to the same species. Such an exchange can be natural, occurring by chance encounters among subgroups; it can also be intentional when initiated by man. Natural exchanges abound in the plant kingdom. They are favored by winds, waters and tectonic movements which disperse pollen to receptive stigmas. Examples of trait mixing initiated by man include the coupling of different types of domesticated species – dogs, cats, cows, chickens or ducks – to accentuate certain traits in the progeny. Pastoralists in Africa and elsewhere have practiced this since the origins of animal domestication. In Tshiluba, a type within a species is called '*diminu*'. The practice of coupling different types is called '*kushinta*', with the connotation of modifying ('*kushintulula*') for the better. Loaning and borrowing of roosters, ducks, goats, bulls, and rams for this purpose are common practice among African farmers.

The human species counts various types that dispersion and geography have consolidated over millennia since the emergence of *Homo sapiens*. Taxonomists, for whatever reason, felt the need to freeze these types or varying ensembles of them under the term 'race'. But, as we shall see later, the term, which was intended to be scientific, has been perverted over the centuries by ideological connotations. Sociologically, the word lumps together individuals and groups of individuals of the vast humanity around 4 or 5 aggregates of traits, making abstraction of alternate combinations of traits. Taxonomists may have initially coined it for comparison purposes, but it ended up being used to evaluate groups of people, to glorify some groups and denigrate others. Today, the word is so full of a long history of social malfeasance or battles that one cannot utter it without prejudice, pride or pain, conscious or unconscious.

Of course, human types are recognizable by their immediately apparent physical features or combinations thereof, such as skin color, nose width, lip thickness, or hair texture. The fact remains that human types are difficult to define because there are various gradients and mosaics of traits of them. In sub-Saharan Africa, the same taxonomists have recognized Bantu, Nilotic, or Sudanese types among peoples with variably 'melanized' skin (the so-called 'Blacks'). Within these types, one can also perceive subtypes. For example, among the Bantu of my native country, the Democratic Republic of Congo (DRC), one can, if one dwells on it, perceive subtypes *'Mukongo'*, *'Muluba'*, *'Mushi'* or *'Muyaka'*, by physical features and mannerisms. However, at this level, the distinctions are usually imprecise, often hazardous, occasionally arbitrary, and sometimes erroneous. These types and subtypes result from the consolidation of traits by a reproduction restricted to a given geographical or cultural space. This differentiation can occur even among peoples of common origin regionally separated for a few centuries.[3] Predictably, it gets diluted as this reproductive space expands and various mixes take place within it. This is the case in the non-ethnic spaces represented by urban centers.

From their understanding of heredity, have Africans developed strategies to improve or strengthen particular traits to the detriment of others? A careful reading of the past recorded in ethnological documents indicates that this was indeed the case.[4] Among Africans (as among other human groups and even animals), a grain of eugenics underlies any reproductive choice: healthy, vigorous, handsome or beautiful men and women, have always been favored as potential sexual partners: they were noticed, coveted, and sought after. In some ethnic groups of the Kasai province of the DRC, when a young man aspired to the hand of a girl, his sisters were entrusted with the task of judging her capacity for domestic and childbirth labor. During communal baths, they were to evaluate her body proportions, breasts, thighs, calves, pelvis, and posterior.

Moreover, Africans commonly refrain from certain couplings that can lead to the 'deterioration' of their lineage, to recurrent morbidity and mortality within them. Endogamy is one of the recognized sources of deterioration. It is therefore absolutely forbidden in many ethnic groups. For the Baluba people, this prohibition applies to the entire clan. A clan can group dozens of villages of common ancestry, sometimes spreading over ten generations and counting several thousand souls. Marriage between members of the same clan is considered incestuous and cursed. It is thus prescribed to anyone contemplating a matrimonial project to seek the eventual spouse in distant clans. In urban centers where various tribes live side by side, it is crucial for a man to inquire about the clan of the woman that his heart desires, before singing her a serenade and asking for her hand, lest he breaks the sacred taboo of the extended incest. In any case, family endorsement of any possible alliance is subject to careful verification of compliance with the prohibition. However, in some DRC's tribes, such as the Bakongo tribe, marriage between cousins, even first cousins, is authorized.[5] The practice may not be so widespread since, in my search of the literature on the subject, I didn't come across any report of a genetic or congenital pathologies attributable to the practice of familial endogamy in the tribe.

In addition to absolute prohibitions, there are strong recommendations aimed at preventing unintended occurrences in the offspring. For example, to a fair complexion person, it is recommended to take a spouse of a darker complexion for fear of engendering an albino child.[6] Very often, a precedent of albinism or simple freckles ('*nsakamuabi*' in Tshiluba, '*lubela*' in

Kikongo) within a family is a cause of deep circumspection when considering taking a spouse in such a family. I still remember with amusement the panic that gripped my mother when I told her about the albinism of the older brother of a beautiful girl whom my older brother was considering for marriage. "*Abu kabundengi!* (No way, I'll let that thing touch me!)", she screamed. With this scream, she vetoed any possibility of an alliance between my brother and the girl.

A similar circumspection is exercised, albeit to a lesser degree, when one is considering taking a marriage partner outside one's tribal boundaries, but this time out of concern for cultural continuity rather than procreative health. The consented assimilation of one partner – the woman in general – into the culture of the other partner (language and customs) alleviates somewhat this concern.

These proscriptions, prescriptions or recommendations constitute evidence of empirical and approximate knowledge of the biology of heredity. This knowledge was assumed, exploited and, to some extent, manipulated to secure the future. In my view, they were scientific in the broadest sense of the word.

1.2. Heredity explained

THIS BIOLOGY WAS LIMITED to the mechanics of reproduction. Through 'practice and observation', Africans have always known that insemination of a female by a male is necessary to reproduce and perpetuate life. The existence of a vocabulary describing reproductive organs and processes testifies to this knowledge. In the Tshiluba lexicon, for example, are found words such as '*nsapu*' (scrotum), '*kamuma ka mikuji*' (ovaries), '*mpaya wa muana*' (uterus), '*nkishabu*' (placenta), '*muoko*' (umbilical cord), '*kuluma*' (to inseminate), '*kuimita*' (to become pregnant). However, it is not sure that the notion of the meeting of two seeds, that of a woman and that of a man, as the initiating event of reproduction has been so widespread in popular science. Instead, Africans referred to conception as '*man's water*' ejected during sexual intercourse mixing with '*woman's blood*' in the womb to form the fetus. They knew that the testicles bore the essence of manhood, that menstruation indicated a woman's reproductive maturity, and that their cessation after sexual intercourse with a man inaugurated pregnancy. This knowledge covered the biological process via ad hoc organs; it did not explain the heredity of traits. The explanation was of another order: it was spiritual.

In traditional Africa – i.e., that had not yet been subjected to the explanatory schemes of imported religions such as Islam or Christianity – heredity found its rationality through the visions that the peoples of this cultural space had developed to give meaning to their existence; that is, through their justifying cosmogonies. With a few nuances, these cosmogonies affirmed the existence of two worlds: the visible and the invisible. Beings of flesh inhabited the visible world; God, spirits, and the ancestors, the invisible one. These two worlds are in permanent communion. In fact, the visible world is nothing more than a reflection, better, a faithful expression of the wills or actions of the invisible world, of God in person, the spirits, and the ancestors (in accordance with the parcels of power granted them by God). Is it not said in Tshiluba: "*Kulela kakuena ku makanda, anu Nzambi wa Kulu wa kuela lupemba.*" [To have a child is not in the power of the human being but of the God of Heavens when he deigns to mark him with kaolin (to bless him)]? It is in this intangible world that the impulses of life originate before they become incarnate.

Thus, for the traditional African, heredity is not a natural law; it is a decision of the spirit world made concrete. It is the reincarnation of a parent or a deceased ancestor. This incarnation can only take place within the extended family. It can be

Chapter 1: Concepts of heredity in African traditions

recognized at conception, at birth, and during the child's growth, by physical or behavioral similarities with the deceased. According to the belief, the time between death and reincarnation is indefinite. It can occur long after death; in this case, only the collective memory of the family and the clan can recognize it. It can happen shortly after death and is immediately recognized and honored. In this context, I would like to mention two death cases that I closely witnessed: that of KTG, a brother of my wife's, in the 1980s, and, more recently, that of BKJ, a friend's wife. In the weeks that followed their passing was born, from the widow of the former, a son and, from the daughter-in-law of the latter, a daughter, who disturbingly resembled the deceased parents. As it should, the babies received their names. The occasional occurrences of such coincidences reinforce the belief in heredity by reincarnation and bring it closer to a scientific theory.

The very naming of a child means the passing on of spiritual as well as physical and temperamental traits. In African families, it is not uncommon to hear of strange resemblances between a child and his or her living or deceased namesake. In the DRC national languages, the word for homonym is '*ndoyi*' in Lingala and Kikongo, '*majina*' in Kiswahili, '*shakena*' in Tshiluba. It certifies a profound and organic twinning between its bearers. '*Ndoyis*' commune to the point that harm done to one of them is psychologically and physically felt by the other. Therefore, one must exercise caution in the presence of namesakes. Wishful thinking, superstitious belief, or empirical observation? Take your pick. The assertion nonetheless reflects an attempt to explain the phenomenon of heredity.

Another form of transmission is that of governing or professional power. For power is an inheritance of the spiritual world. It can be transmitted within the lineage (by birth) or outside it (by initiation). In the latter case, transmission is mediated by a person endowed with this specific power. In this category, we find hereditary succession on thrones of kings and chiefs, as well as membership to various professional brotherhoods of mystics, priests, mediums, therapists, hunters, soldiers, sailors or metallurgists, etc.

I once witnessed a folk ritual which was intended to convey the character traits of a woman to a young girl. The occasion was festive; the woman stuck her lips against the child's and poured her saliva into the child's mouth saying, *"Wamfuana!"* [May you resemble me!]. The gesture was intended to be a favor and an endowment. Research on the subject has led me to believe that this practice was once widespread across cultures and eras.[7] In all cases, this sharing of biological fluids symbolized (or actualized) intimacy, a pact, or the transfer of life force from an elder to a youth, from an ascendant to a descendant.[8]

Life force can be conferred; it can also forcibly or maliciously be acquired. In Africa, we often hear about esoteric rituals of willful acquisition of such a force through the manducation, real or mystical, of the flesh of a dead person who was considered powerful in life. The secret behind the burial of the high personalities of traditional societies is intended to prevent the profanation of their graves for this purpose. When one thinks about it, the practice resembles, *mutatis mutandis*, the Catholic Church rite of the Eucharist and its worship of bones, garments or other articles having belonged to persons proclaimed or presumed to be saints.

1.3. Morbid heredity

THERE IS NO DOUBT that ancient Africans did not fail to notice that some deadly diseases occurred in families through generations. Causes of these diseases included either a relentless curse stemming from an ancestral sin or the malice of a spirit (called *'ogbanje'* by Igbos people of Nigeria) that takes flesh for a while only to depart shortly after that, thus subjecting the parents to repeated ordeals.

Most often, one could reverse the curse by paying a tribute (immolation of a domestic beast or poultry) accompanied by supplications to God and the ancestors. Some tribes practiced a more or less elaborate rite aimed at interrupting the incarnation-disincarnation cycle: it consisted of a slight mutilation of the newborn's body, making it undesirable for the spirit world. To that end, Igbo people used to sever at the metacarpus the index finger of an *'ogbanje'* child's hand.[9] As a result of such a propitiation ceremony, subsequent children were sometimes allowed (or agreed) to remain alive. Incidentally, among the Baluba peoples of central DRC, my name 'Majambu', which means 'sepulcher', is attributed to the child who has put an end to such a series of childhood mortality. As for me, I inherited the name; I didn't deserve it.

A hereditary disease which used to be fatal for children under five years of age had as symptoms atrocious pain in the joints, gross deformity of the skull, unilateral swelling of the abdomen, epidermal pallor and generalized frailty. The disease had various names in West African societies: *'Chwechweechwee'* in the Ga of Ghana, *'Ajuhoi'* in the Igbo and *'Amisani'* in the Haussa of Nigeria, *'Itaagmi'* in the Bassari of Togo, *'Adep'* in the Banyangi of Cameroon.[10] Modern science has demystified it; it has called it sickle cell disease (SCD) or drepanocytosis.

In the DRC, the prevalence of sickling trait is about 30%; the prevalence of the disease per se is about 2%. However, I have not found any archived oral or written documentation of this region that indicates an early identification of the disease by a specific name. It is hard to believe that the typical symptoms of the disease could have escaped the attention of parents and therapists, to the point that they did not seek to name them. On this onomastic vacuum, I interviewed some Congolese doctors who had professed for a long time in the Congo. Without being certain, one of them, a surgeon, claimed to have heard of the '*bokono ya kibeka*', Lingala for 'spleen disease', referring to splenomegaly (enlargement of the spleen) which is a mark of SCD. A pediatrician offered me this reasonable explanation: It may be that, long ago in this part of Africa, SCD children died shortly after birth, before the symptoms of the disease could become apparent. In southwestern and central DRC there was an ancient rite called '*mbombo*' which was intended to treat cases of recurrent miscarriages or perinatal mortality.[11] It possible that the rite also applied to cases of deaths of SCD children.

Hereditary or not, illness means disability or pain; it calls for relief by therapeutic intervention. In the pre-scientific world of traditional Africa, hereditary diseases were interpreted as an ancestral curse which must be expunged before health can be restored in the lineage. Therapies based on symbols, prayers, ablutions and herbal concoctions were applied, often concurrently. This is still the prevalent mode of operation among the rural Yoruba of Nigeria when confronted with SCD.[12]

However, in ancient Africa, a physical trait that ran through generations in a given population with no apparent morbidity, was sometimes considered normal and even desirable by the people, while other peoples would take it for an anomaly. The goiter was a case in point. This enlargement of the anterior part of the neck raged in Ubangi-Uele of the Belgian Congo at the beginning of the last century.[13] The natives 'adopted' this deformity as a sign of feminine beauty, in spite of the slight mental retardation (cretinism) that it sometimes caused. We now know that the enlargement affected the thyroid gland following a chronic dietary deficiency of sodium iodide. Addition of this salt to the diet quickly led to the disappearance of the anomaly, demonstrating that it was not hereditary, but acquired. In the same manner, 'Lobster People' of the Vadoma tribe in the Kalahari Desert were quite comfortable with their feet made of two toes, the largest and smallest, bulky, padded, and claw-shaped. Ectrodactyly is the scientific term for the deformity. It is due to a genetic mutation. Tribal endogamy contributed to its propagation.

1.4. Unusual births

IN TRADITIONAL AFRICA, every birth that was out of the ordinary carried a message. In Baluba culture, there are several recognized cases of such births: '*Mapasa*' (twins), '*Musuamba*' (born after twins), '*Ntumba*' (conceived without prior menstruation), '*Tshiowa*' (whose gestation is accompanied by menstruation), '*Tshiela*' (who leaves the mother's womb feet first), '*Ndomba*' (who present a hand before exiting the womb).[14]

Among these births, that of twins was unquestionably the most significant. Because of its singularity, it usually aroused fear of a spell if one dared contradict twins' desires, an anxiety which, in the extreme, could lead to preventive infanticide. A British anthropologist, who lived with a Bakongo tribe at the beginning of the last century, wrote about it:

> Because of the extra trouble they entail, the Lower Congo women do not take kindly to twins, hence it is the general practice to starve one of them. When a twin is thus starved, or dies a natural death, a piece of wood is roughly carved to represent a child, and it is put with the living twin that it may not feel lonely. Should the second child die the image is buried with it. The corpse of a twin is placed on leaves and covered with a white cloth, and is buried at the crossroads like a suicide, or a man killed by lightning. It is regarded as a hateful thing, and is buried in the most dishonorable of all graves.[15]

The traits of 'unusual' children are not hereditary; they are episodic deviations from the norm, i.e., they are exceptions. Traditional societies had instituted more or less elaborate and sometimes recurrent ceremonies to appease such children. However, this acceptance adds compassion to apprehension. It also applies to all forms of congenital anomalies. Examples include dwarfism (short stature), cleft lip, hyperkyphosis (hump), trisomy 21 (mongolism), polydactyly (supernumerary finger or toe) and syndactyly (finger or toe fusion or palmure). Two Baluba proverbs encapsulate this acceptance: *Walela kuimanshi, nansha ka mutu mpandu* [Do not reject the baby you gave birth to, even if it has a split head]; *Kuseki mulema, ne upange kufua disu, watshibuka mukolo* [Do not make fun of the crippled lest you lose an eye or break a leg].

In summary, traditional African societies were well aware of the phenomenon of heredity. By pure empiricism, they understood its main biological mechanisms. They even theorized about it by drawing on the philosophical and religious postulates of their cultures. Traditional theories of heredity possessed an inherent coherence and logic. The incorporation of disembodied wills into these theories ensured this coherence. Moreover, this is not unique to the African world: at all time, in all cultures, the call to spiritual hypotheses has been part of the search for meaning in life and the Universe. In any case, these theories have allowed African societies to manage their reality, to orient themselves by making useful choices that have ensured their survival for millennia. By definition, any theory has a margin of uncertainty. Only careful observation of the consequences of the said theory can allow its margin to be circumscribed and consolidated. Traditional African societies have preserved these theories because, in one way or another, they served them well, from a practical, therapeutic, and psychological points of view. They have given rise to a wealth of knowledge.[16] This knowledge, unfortunately, risks disappearing before the indifferent eyes of converts to modernism. They reject or question the objectivity of the predictions derived from these theories because they live outside this reality, having

Chapter 1: Concepts of heredity in African traditions

forged another one more reassuring, more practical, but just as burdened by its own uncertainties. Indeed, the use of the scientific vocabulary does not free man from his existential anxieties. It rearranges his perception of reality, grants him a certain mastery over it, but leaves him just as dumbfounded by the mystery of life and the Universe, as apprehensive and vulnerable in the face of his inescapable finitude.

Chapter 2

From Heredity to Genetics

IN ITS ORIGINAL ETYMOLOGY, the word science means a sought-after accumulation of knowledge, born of the desire to understand with the goal of better managing the future. Until proven to the contrary, it is a purely human activity, as animals are guided only by the instincts enshrined in their biology. Observation of facts is the mother of science; her daughter is the arrangement of these facts in schemes that are both palatable to reason and open to useful predictions. Since his emergence, man has always been a scientist, even though, for long, his science has accommodated itself of divine and spiritualist postulates.

Thus, to say that science started only three centuries ago is a redefinition of the term. What preceded it is now called protoscience. The new definition has purged science of intangible and uncontrollable wills, and has confined it to the observable, the controllable, and then verifiable. It has rejected any speculation that does not meet these criteria. Theretofore, science had nothing to do with gods or spirits, nor with theological and philosophical proclamations. It is, or at least must strive to be, material, impassible, and objective. Like religion and philosophy, it allows for abstract speculations, but unlike them, it demands that these speculations lead to experimental verification. Coincidentally, in the field of heredity, it is a *'man of God'* who first applied these new norms to the phenomenon and turned it into a science called genetics.

2.1. The monk's garden and the great theory

Gregor Mendel (1822-1884), a Czech Catholic monk, was an impassioned experimental botanist. In a garden around his monastery in Brno, he cultivated a large variety of pea plants – 30,000, reportedly – and systematically studied their characteristics, plant sizes and pea colors in particular. Each variety of pea plant is capable of self-fertilization and cross-fertilization with other ones. Self-fertilization over several generations produces pure lines of plants. Mendel had developed such lines. When he crossed a short line with a tall one, plants resulting from this crossing (F1 generation) were all tall. He concluded that 'tallness' was a dominant trait. But when he allowed F1 plants to self-fertilize, the new generation F2 plants were 75% tall and 25% short. From this distribution, he put forth the following hypothesis: (i) in each pure line, the trait 'size' was determined by a pair of identical factors of 'shortness' (ss) or 'tallness' (TT); (ii) in the F1 generation, each line contributed half of its pair to form a mixed pair (sT), but since the 'T' factor was dominant and the 's', weak (recessive), all plants of the generation were tall; (iii) to produce plants of generation F2, the factors separated and recombined randomly in ss, sT, and TT, in the proportions of 25%, 50%, and 25% respectively, giving rise to 25% of short plants (ss) and 75% of tall plants (sT and

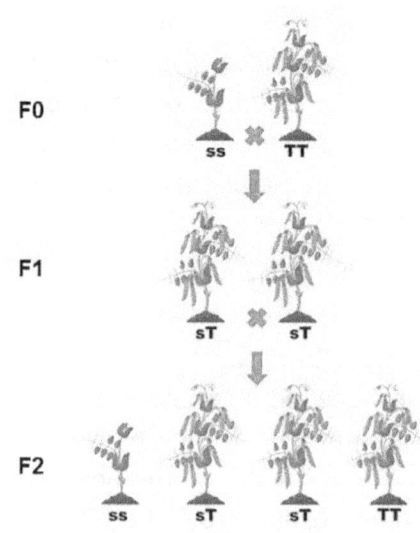

Figure 2.1. : *Mendel's experiment*

Chapter 2: From heredity to genetics

TT) (Figure 2.1). Mendel confirmed these laws by examining other traits, such as pea color as well as pod, leaf, or seed shape.

Today, we know the nature of these pairs of factors, the seeds that carry a single copy of them, and how they combine during sexual reproduction. I will describe them later in this section. For his quantification of traits and *'mathematization'* of the laws of their transmission, Gregor Mendel has been crowned Father of Genetics. However, his discovery was not known until 40 years after its publication,[17] and 20 years after his death, drowned as it was in the effervescence of biological observations collected during this period and the theories that sprang from them.

The most brilliant of these theories was undoubtedly that of the evolution of species, articulated – not conceived – by the English naturalist Charles Darwin (1809-1882).[18] It stipulated that the traits of species were subject to natural selection which allowed preferential transmission from generation to generation of those traits that fostered the survival of the species. In other words, Nature continuously subjected species to various forms of pressure, so that those that lacked the traits allowing them to sustain these pressures perished, and those who possessed them survived.

It has been a century and a half since Darwin published his theory. The theory has been fruitful in many aspects of biological sciences. It has constituted an encompassing framework under which biological phenomena are scrutinized. A unifying theory of sort. Amendments have been to it through the years; it has withstood the test of relevance by and large. As of now, no other theory can better explain the emergence and evolution of life on Earth.

However, the application of the theory to social issues has been prone to misuses and abuses. The theory has been gleefully distorted by 'soft' scientists (e.g., sociologists, psychologists, etc.), pseudoscientists, and amateur scientists to gauge and rank individuals and societies, and to justify all sorts of social injustices, even massive crimes. We will expand on these distortions in chapter 4.

A keen observer of natural similarities and differences, Darwin also speculated that species did not appear readymade, but that they descended from ancient species as a result of differentiation and consolidation of distinctive traits. Geographic and reproductive isolation contributed to these differences. For Darwin, species resembled the branches of a *'Tree of Life'*, the more proximal ones being more similar, the more distal ones more dissimilar. Thus, because they share most physical similarities, men and chimpanzees were considered close relatives who descended from a long-extinct common ancestor. In the last few decades, molecular evidence has amply corroborated this hypothesis.

Some say that Darwin's view on the evolution of species is no longer a theory, it is a fact. Yet for more than 150 years, despite the sum of paleontological and molecular evidence in support of it, this view continues to offend the sensitivity of believers in the biblical legend of creation in seven days. Sadly, it has also served as an argument for those who try to rank individuals or groups of individuals within species, and even within the human species, to explain social inequalities and injustices through the supreme rule of *'survival of the fittest'*.

2.2. Blueprints to bequeath

WHEN MENDEL AND DARWIN left this world, the microscope had already been invented. It permitted to see the most elementary forms of life, capable of living independently and of reproducing themselves: microbes. It had also shown that tissues of complex living things – plants, fish, insects, animals, etc. – were made up

Chapter 2: From heredity to genetics

of a myriad of small units called *cells*, more or less different, and attached to each other. Cells would multiply before they took the shapes and textures appropriate for their particular tasks. These microscopic observations led to the conclusion that cells assembled into tissues, tissues into organs, and organs into organisms. At this point, the program that guided these differentiations and assemblies was completely unknown.

Cells are miniscule bags bounded by a membrane and containing the essential ingredients for their survival, including the life blueprints on which are inscribed protocols for the making and transformation of cellular structures and ingredients. When a cell is made of a more or less homogeneous interior, it is called prokaryotic (e.g., bacteria). When its interior consists of two major compartments, a *cytoplasm* and a *nucleus*, it is called eukaryotic (e.g., plant, animal, and human cells) (Figure 2.2a). The cytoplasm is a 'plant' that manufactures and processes the ingredients essential to cellular survival and functions, including proteins, sugars, and fats. The nucleus is a 'vault' within which are stored life blueprints. On necessity, cells transcribe the blueprints into temporary copies that travel to the cytoplasm where they guide the production of the 'factory workers' called proteins. Of these proteins, *enzymes* are essential for life within the limits of time allotted to each cell or organism: they speed up manufacturing and processing reactions that would otherwise take too long to complete. Enzymes are true catalysts of life.

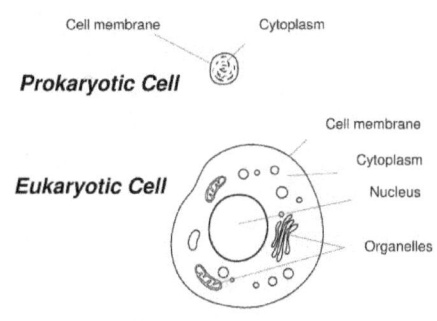

Figure 2.2a.
A pro- and eu- karyotic cell

THE COLORS OF GENES

Blueprints reside in the nucleus as small elongate packages called *chromosomes*. Most cells which make up the human body – for henceforth my narrative will focus on man – contains 46 chromosomes, in fact, 23 pairs of similar (but not identical) chromosomes. These are called *somatic cells*. Cells implicated in sexual reproduction (*gametes*) contain only one copy of each chromosome for a total of 23. Hence, somatic cells are said to be *diploid*, and gametes, *haploid*. Of the 23 chromosome pairs of somatic cells, 22 are called *autosomes*; they are numbered 1 to 22, from longest to shortest. The remaining pair is called 'sexual'; it is identified by the letters XX in women, and XY in men.

One property typical of a somatic cell is its ability to replicate itself. To do this, it doubles all its chromosomes, equally amasses them to two extremities, shares the cytoplasm and the membrane around these clusters, and splits the latter, thus forming two identical cells. This type of cell division, called *mitosis*, is the primary way of transmitting the blueprints of life from a mother cell to two daughter cells. It helps grow tissues and organs to their optimal sizes, to renew, maintain, and repair them.

In the reproductive organs – the ovaries in women, and testicles in men – mitosis of germ cells progresses to *meiosis*, or the formation of haploid gametes – ova (eggs) in women, and spermatozoa (sperm) in men. When a man inseminates a woman and a sperm penetrates an egg (i.e., fertilizes it), the fusion leads to the formation of a *zygote*, a reconstituted diploid somatic cell which combines the chromosomal halves of the two partners. Together, these chromosomes will direct cellular multiplication and differentiation, leading the zygote to develop, successively into an embryo, a fetus, a baby, a boy or a girl, a man or a woman. This, in short, is the process of human reproduction; this, in essence, is the mechanism of human heredity.

This transfer of the blueprints of life by fusion of the two parental gametes serves to reproduce the generic organism and

Chapter 2: From heredity to genetics

propagate the human species. Sexual reproduction has other advantages: it leads to the mixing or sharing of biological capabilities, thus mitigating weaknesses and accentuating strengths for survival and procreation; it also allows greater variability of these capabilities, an adaptive advantage in the face of environmental changes. I will discuss this concept in more detail in later sections dealing with the chemistry of genetic variations. In short, the combination of the chromosomal heritages of different people of the same species, called *heterozygosity*, is a biological advantage; its loss is usually a bad omen.

At the dawn of the 20th century, it had become clear that carbon, hydrogen, oxygen, and nitrogen represented the atoms of all living things. Life, therefore, is nothing more than molecules based on these atoms, assembled, organized and dedicated to various and distinct tasks of building, destroying, purifying, repairing and perpetuating. To study the mechanics of life is to unravel the chemistry of life, to study its biochemistry. To understand heredity is to explain the biochemical reactions that allow organisms to reproduce and perpetuate themselves.

When cytologists (scientists studying the architecture and the functioning of cells) described the stages of mitosis, it was clear to them that chromosomes were the most visible cellular material transmitted from mother cells to daughter cells. From this observation, they inferred that chromosomes contained the molecules that ensured heredity. Biochemists extracted this material from the cell nuclei and determined its composition: it consisted of long acidic filaments called *nucleic acids* which was made of random sequences of four similar units called *nucleotides*. Each unit carried a phosphate group, a sugar, and a flat, closed multangular structure called *base*, identified by the letter A, C, G, or T. Bases differentiated nucleotides; phosphates attach them to each other; the sugars were of two types: ribose and deoxyribose. Nucleic acids composed of deoxyribose were named *deoxyribonucleic acids* (DNA); and those composed of ribose, *ribonucleic acids* (RNA). Biochemists concluded that chromosomes consisted primarily of DNA coated with of a few proteins and RNA, that *DNA was the molecule of heredity*.

The next challenge facing these chemists of life was to find out how these DNA filaments could replicate and be transmitted from cell to cell, from parent to child. The path to solving this challenge was the observation of curious symmetries in the chemical composition of DNA. Indeed, biochemists noted that the sum of A bases was equal to that of T bases, that of C bases, equal to that of G bases. Moreover, when chemists subjected DNA to X-ray examination, on a sensitive film that captured its image, they saw a circular pattern punctuated by concentric dark spots at equal distances from each other, like the shadow of a serpentine illuminated from the center over its length.

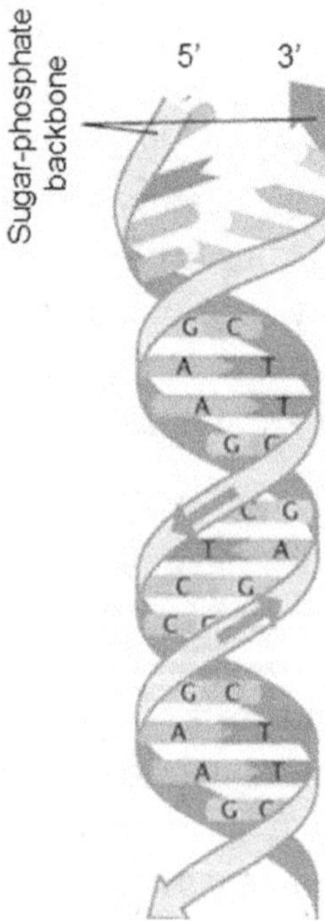

These were the data in the mid-20th century. What to make of them? What to deduce from them? As recounted by Horace F. Judson in his book evocatively entitled "*The Eighth Day of Creation*",[19] two young researchers from Cambridge University, the American James Watson and the British Francis Crick (1916-2004), appeared on the scene. In a flash of genius, they took the data and connected the dots. They conceived the *DNA double helix* model (Figure 2.2b). Through this simple act of imagination published on a page and a half,[20] they illuminated the mechanisms of life's replication, giving birth to a science now

Figure 2.2b :
The Double Helix

known as molecular genetics. The discovery earned them the Nobel Prize in 1962, which they shared with Maurice Wilkins (2016-2004) who conducted the critical X-ray imaging of DNA.

In Watson and Crick's vision, DNA consists of two antiparallel [parallel, but growing in opposite directions (*see* arrows in Figure 2.2b)] strands of nucleotides twisted in serpentine.

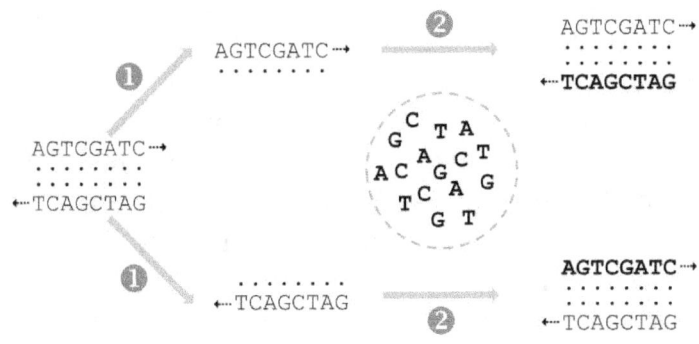

Figure 2.2c. *Replication of the double helix*

The bases of these strands face each other: A faces T, and C faces G. Thus, A is said to be complementary to T, and C, to G. For DNA duplication (Figure 2.2c), the double helix untwists itself lengthwise (step 1). Each strand then serves as a matrix for the manufacture of a brother strand, by placing complementary bases drawn from a nucleotide reservoir (step 2, in bold letters). And there you have it: a replicated double helix!

Simple? Not so simple! The process involves the programmed participation of a good number of specific enzymes, which place each nucleotide in the right place, carefully check the placements, remove and correct them in case of error. The quality control sometimes fails and leads to the introduction of changes in the replicated nucleotide sequence. The latter can be inconsequential, beneficial, or pathogenic. They can be transmitted to daughter cells (somatic changes) within a living organism, or to progeny (germline or hereditary changes), as discussed below.

Chromosomal DNA resides in cellular nuclei: it is called nuclear. DNA also exists in *mitochondria*; these are small organelles scattered by the hundreds in the cytoplasm of all cells. They originated from microbes which colonized the eukaryotic cell long ago during evolution. They serve as cellular furnaces, producing ATP (adenosine triphosphate), the chemical energy which fuels biochemical reactions. Although found in all cells, men's and women's alike, mitochondria and their DNA are transmitted from mother to daughter only, not from mother to son, not from father to son, not from father to daughter. The importance of this mode of transmission will be evident when I address the question of where in the world the first woman appeared.

Cells can replicate DNA. They also transcribe it, according to the same rules of complementarity of bases, but in single strands and with ribose-sugared nucleotides (instead of deoxyribose) and bases A, C, G, and U (instead of T). *Transcription* generates RNAs, which represents temporary repositories of the information contained in DNA. Unlike that of DNA, their amount and variety constantly change through the lifetime of a cell. Produced in the nucleus, they travel to the cytoplasm where they act as matrices for *translation*, i.e., the manufacture of proteins, the real body pillars and cellular workers.

Untranslated RNAs associate with proteins to build organelles dedicated to various functions (e.g., ribosomes making proteins) and to regulate various cellular processes (i.e., when, how, and how much proteins are made).

Thus, the normal flow of genetic information consists of a loop (replication) following by a transfer in two successive stag-

Chapter 2: From heredity to genetics

es (transcription, and then translation, as illustrated in Figure 2.2d.

Figure 2.2d. *The flow of genetic information*

2.3. The bricks of the genome

As mentioned above, DNA is made up of long chains of four types of nucleotides – A, C, G, and T – randomly succeeding each other. However, not all DNA qualifies as a gene. What is a gene? In its restricted structural meaning, a gene is a segment of DNA containing information for a precise task to be performed after it has been converted into an RNA and or a protein. In its broader sense, the definition also includes parts of DNA, in or out of the gene proper, which control where, when and how much of the information is expressed. Thus only when it specifies a cellular function does a string of successive nucleotides merit the title of 'gene'. All the genes of a cell collectively form its *genome*.

In the late 1970s, two chemists, Fred Sanger (1918–2013), the British from Cambridge University, and Walter Gilbert, the American from Harvard University, developed two different methods of determining the nucleotide sequence (*sequencing*) of DNA. For developing these technologies, they were awarded the 1980 Nobel Prize for Chemistry. Soon, engineers automated the more user-friendly Sanger's method, making sequencing easy and general. In the meantime, computer capacities for data storage and analysis grew exponentially.

Two Americans, Francis Collins of the National Institute of Health (NIH) (a government agency) and Craig Venter of Celera Genomics (a private company), along with their rival battalions of researchers, decided to elucidate the complete nucleotide sequence of the human genome. In year zero of the current millennium, these 'Indiana Jones' of biological research met the challenge: they opened the vault; they revealed to humanity the chemistry of its emergence and permanence, in short, of its heredity.

The human genome contains about 3 billion pairs of successive bases, but only 30 thousand genes. The average length of a gene being about ten thousand base pairs, the share of DNA needed for human life would be only 300 million base pairs, or 10% of the genome. What about the extra DNA? What is it for? Is it truly useless to life? Some called it *'junk DNA'*, dumpster or garbage can. Others prefer to see it as a memory and an experimental ground of evolution. Not only this DNA contains traces of our past encounters with plants, bacteria, and viruses, but it also undergoes rearrangements, somersaults, and contortions, as though it was exploring or anticipating other life possibilities without threatening life today. Speculation? Yes, but if one contemplates the long train of our evolutionary past, this hypothesis does not seem so much off the track.

When we look at the alignment of genes along chromosomes in different species – say, mice and humans –, we see impressive homologies across entire segments. Synteny is the name given to this observation. Scientists interpret it to mean that that the process of species apparitions involved a rearrangement of entire chromosomal modules; that, in the distant past, there existed an ancestor common to humans and mice

Chapter 2: From heredity to genetics

whose genome segments underwent reorganizations to generate the precursor species of humans and mice. This ancestor presumably lived 80 million years ago. The evolution of species thus corresponds to the evolution of genomic modules, their restructuring and differential alterations, which ultimately prohibit the mixing of genomes, impeding joint transmission of a unified genetic heritage. Thus, species are set apart when their genomes 'divorce' i.e., become incompatible.

This combinatorial game, evolution also plays it in the creation and diversification of capacities between and within species, especially within complex species such as animals. To succeed in the game, Nature has forged genes into functional modules. Thus, in humans, the information contained in a gene is made up of informative segments named *exons*, interrupted by non-informative ones called *introns* as illustrated Figure 2.4 (upper line).

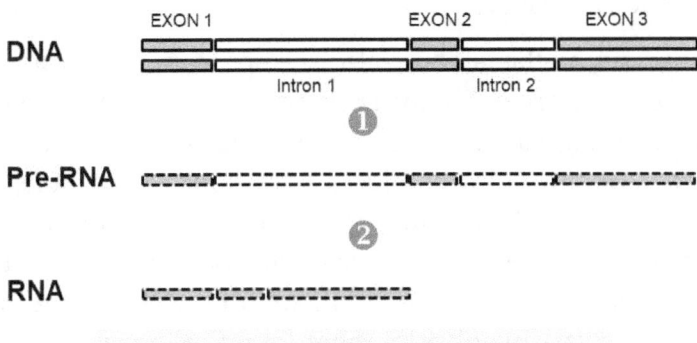

Figure 2.4. *Spicing of exonic segments*

When transcription of a gene into RNA takes place, it does so initially over its entire length of the gene (middle line). Processing of this primary RNA (pre-RNA, Figure 2.4, middle line) involves removal and destruction of segments derived from introns, with concomitant orderly abuttal of those derived from exons (Figure 2.4, bottom line), forming a functional

RNA, to be translated into proteins or to build organelles.

This conversion of pre-RNA into RNA is called *splicing*. It is an orderly process, obeying to rules which are specific, but not absolute. Indeed, during gene transcription in some tissues, *alternate splicing* sometimes occurs, inserting introns or skipping exons, and altering the makeup of the final products. Alternate RNAs may exhibit different longevity within cells or direct the manufacture of different proteins with distinct functions.

Moreover evolution has often combined exons from different genes to create new genes that would dictate new information for new tasks. *Evolution obeys the law of parsimony*. It recuperates, recycles and rearranges modules of DNA to create diversity. It borrows, acquires, and sometimes combines DNA from various sources. For example, it is estimated that 10% of our genes may have come from viruses. Remember mitochondria? These bacteria which colonized ancient eukaryotic cells and brought their DNA along, add a supplementary substratum for genetic variability?

Diversity in the genome also springs from occasional, more or less circumscribed changes known by the generic name of *variations* or *polymorphisms*. These include translocations or duplications of chromosome pieces, insertions or deletions of nucleotides, above all and in large numbers, unitary changes of nucleotides captured by the acronym SNP (for *single nucleotide polymorphism*). There are about 10 million SNPs in the human genome, the vast majority of them found in introns and the defamed junk DNA. In later chapters, I will explain the importance of these variations for understanding the history of man on Earth.

In a population, the same gene can exist in versions differing by a nucleotide or more in their DNA chains. Each version represents an *allele* of the gene. Human somatic cells possess two alleles of each gene, one inherited from the father and the

Chapter 2: From heredity to genetics

other, from the mother. An individual who carries two different alleles of a gene is called *heterozygous* for that gene; he who has two identical copies of an allele is a said to be *homozygous*. A linked combination of variations in a chromosomal region is called *haplotype*. When we imagine the number of nucleotides (3 billion) and the ability of each nucleotide to generate 4 SNPs, the possible combinations of alleles and genetic haplotypes are incommensurable. Diversity is a hallmark of the genome. The total number and variety of alleles in a given population constitute its *genetic pool*.

Variations can occur randomly, at any time, in any cell, due to DNA replication errors, cellular aging mechanisms, ionizing radiation, exposure to toxic chemicals, etc. But, let us emphasize this, *these variations become hereditary only when they occur in the genome of gametes which transmit them to the next generation after sexual reproduction*; and, depending on whether they do not affect, positively or negatively affect survival, their frequencies in subsequent generations will remain constant, increase, or decrease, respectively. Thus, genetic variations, in all their forms and varieties, are the substrates upon which natural selection, acting through various environmental pressures, favors over time the perpetuation of the best-adapted species.

Neutral variations are without any effect on gene expression. In all species, those due to DNA replication errors occur in more or less constant numbers per generation. From differences in variety and number of neutral variations between two related genomes, i.e., two related species, it is possible to estimate the number of years passed since the two genomes diverged, to trace back their history and journeys on planet Earth. Thus neutral variations represent a *molecular clock*. Initial time computations used protein sequence similarities and differences among species. When DNA sequencing became routine, comparative overview of nucleotide variations provides further corroboration and refinement of the calculations.

Any genetic variation defines a *genotype*. In theory, there are as many genotypes as there are variations in a genetic pool. Some variations are linked, i.e., always appear together in a chromosome; others are not, either because they are located on different chromosomes or they can be separated during gamete formation, i.e. during meiosis. I will discuss this further in Chapter 3 when dealing with the signatures of natural selection. A manifestation – an unusual physical trait or a physiological characteristic – caused by a genetic variation is called a *phenotype*. There are genotypes without apparent phenotypes. However, environmental changes can lead to the appearance of novel phenotypes linked to previously silent genotypes. We shall discuss this point in later chapters dealing with melanin genes and vitamin D deficiency in relation to human migrations.

What is a *mutation*? The term simply means 'change'; it is synonymous with variation but generally refers to one that causes a morbid phenotype, otherwise known as pathogenic variation. When they negatively affect survival and fertility, mutations tend to disappear over time due to cleansing by natural selection as those who carry them fail to transmit them to their offspring. Their persistence in a population over time indicates that, somehow, they provided a survival benefit in the evolutionary past of that population.

2.4. The motions of the genome

SO FAR, WE HAVE ONLY DEALT with the structure of the genome, of its multi-colored bricks used to erect those cathedrals we call species. However, the genome is not solely a series of heritable blueprints, carefully guarded and secretly copied in the cell nucleus. There is life around them. Molecular workers – proteins and RNAs – abound and bustle about them. In due course, once or multiple times, these

Chapter 2: From heredity to genetics

workers pull the blueprints out of their chromosomal drawers, unfold, illuminate, and transcribe them, roll them up and store them again in their drawers. The chronology, the diversity, the succession, the combination, and relative abundance of gene expression determine our lives from conception, during growth, development, decline, and death. In biology, everything depends on, and reflects the expression of genes; everything is dependent on the motions of the genome, as genes 'wake up, act, slumber, nap, or sleep', upon the intrinsic guidance and in response to the environment.

Transcription is the first step in gene expression. It takes place in the nucleus. *Promoter* regions, proximally located upstream of the first exons, drive it. Their activity is in turn subject to stimulation or inhibition by other genomic domains occupied by batteries of proteins, the so-called *transcriptional factors*. The cytoplasm manufactures these factors and send them to the nucleus in response to cellular needs.

Translation is the second step in gene expression. It is the manufacture of the ultimate workers of the cells, tissues, organs, or organisms: the proteins. Their activities are dependent on or modulated by various modifications collectively called *post-translational modifications:* e.g., addition of sugars or lipids, tagging with address labels which indicates their final destinations within the cell, chopping to generate fragments possessing distinct activities, or marking with longevity signals which determine how long they will be at work before retirement and destruction.

The cascades and combinations of DNA, RNA, and proteins that trigger and accompany genomic motions take into account and are responsive to the environment surrounding the genome. Indeed, *we are not only a reflection of our genes; we are also the products of our environment.* The latter is the space which determines the results of gene expression, whether this expression grants us survival, health, fertility, disease, or death. Genome and environment are two inextricable realities in biology. It is utterly nonsensical to talk about one without the other. In my opinion, of these two, the environment would be primordial because it has allowed the appearance of the chemicals of life in the primeval soup, of DNA, RNA, and proteins.

Thus, if genetic variations are the substrates of natural selection, the environment is its breeding ground. In fact, it is the ground and the breeder. Without it, life cannot become manifest. The point is of importance in this essay contradicting the undue primacy attributed to genes and genetics in the discourses without nuances of some scientists in search of an admiring audience, and in the imagination of an audience drawn to fatalistic distinctions. As we will see in the following chapter, genetic diversity intertwines with environmental diversity to create myriads of living individualities.

2.5. The fingerprints of environment

THE ENVIRONMENT DOES not experience cataclysmic changes on a daily basis. Bombardments of the Earth by massive meteors, monumental solar eruptions, glaciations that freeze soil and life, are events that occur at geological time intervals. They decimate certain species and spare others more adapted to the new world. In the intervals between such events, surviving genomes consolidate and spread the genetic variations that permitted their survival. Stephen Jay Gould (1941-2002), the renowned Harvard University paleontologist, evolutionary biologist, and science historian called 'Punctuated Equilibrium' this mechanism of brutal selection followed by lengthy and slow affirmation of beneficial traits.[21]

On the other hand, fluctuations in the environment are a constant and commonplace phenomenon. They are multifaceted; they can be physical (e.g., nutrition and climate) or psycho-

Chapter 2: From heredity to genetics

logical (e.g., stress, emotions). Do these fluctuations affect the genome? The answer is yes! We know that they mark specific genes with discrete labels (mostly those called 'methyl groups') to influence their expression. These so-called *epigenetic markings* occur in both somatic and germ cells. Those occurring in germ cells actively modulate the development of the zygote to the embryo, fetus, and adult. They subtly influence the size of organs and the physiological exchanges among them. If the environment remains the same, they can be recapitulated in subsequent generations. Such markings lend support to the famous hypothesis by the British epidemiologist David Barker. He postulated that adult metabolic diseases – obesity or diabetes, high blood pressure – originate at the fetal stage of development.[22] Experimental studies in mice support this hypothesis. In my own laboratory, I have observed that mice born to a mother fed a low-fat diet during pregnancy became diabetic six weeks after birth if fed a high-fat diet; while those born to mother on the latter diet were spared the disease when kept on the same menu. The hypothesis still awaits evidence in humans. *Metaphor aside, there is some truth to the assertion that we are what we eat.*

The main characteristic of epigenetic changes is that they only last for one lifetime; although, the environment being the same, they could reoccur over generations, they are not hereditary in the strict sense of the word. They affect the motions of the genome in the current generation; they are erased in the gamete genomes and reintroduced in the zygote of the next generation, to meet the needs of the new environment.

In summary, of the abundant genetic nomenclature I have just reviewed, please remember this: while supporting life and transmitting it through generations, the genome is variable in thousands of ways, not only in its structure and expression but also in the signatures affixed to it by the environment.

CHAPTER 3

THE BRIEF HISTORY OF THE HUMAN GENOME

3.1. Human life is a wink

THE UNIVERSE HAS EXISTED for 14 billion years. The Milky Way, the galaxy that encircles us, is among the first to emerge from the Big Bang explosion. Our star, the Sun, formed 5 billion years ago, and, half a billion years later, appeared in its surroundings a planet promised to greatness, at least in my view: the Earth. It will take another half a billion years to see the first stuff of life – nucleotides, amino acids – produced in the primordial soup, boiling with gas, water, and soil, and bombarded by sunlight, by visible, ultraviolet, and gamma rays.[23] These elemental materials grouped by chance encounters; they formed molecules, then macromolecules (RNA or DNA) capable of reproducing themselves; these were joined by other macromolecules (proteins) which serve to nurture and preserve this capacity.

Of DNA, RNA, and proteins, which appeared first? No one knows. The 'RNA World' theory stipulates RNA, not DNA, is the first material of heredity.[24] It is founded if the fact that some RNAs can replicate themselves and act as enzymes. It suggests that primitive RNAs were imbued with these dual capacities which, over time, they divided between DNA and proteins.

THE COLORS OF GENES

From productive interactions among these chains arose LUCA, the *Last Universal Common Ancestor*: the primitive cell.

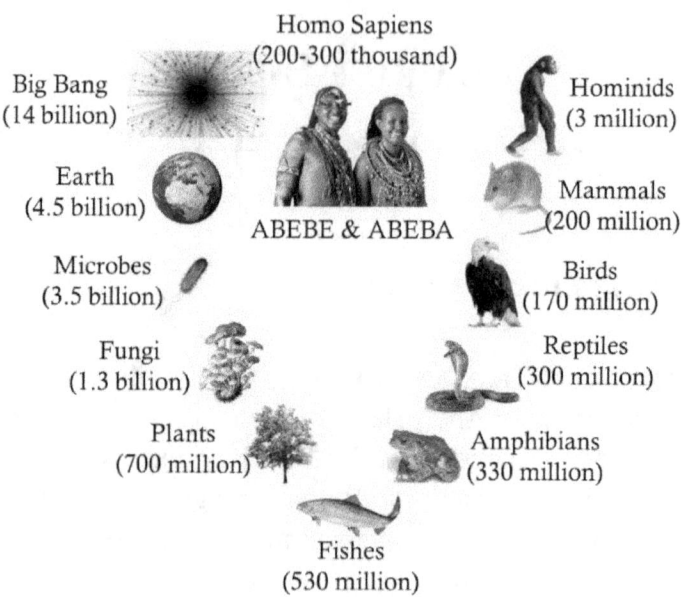

Figure 3.1. *We are so 'recent.'*

From the aggregation of cells, their structural and functional differentiation under the filter of the environment emerged the whole terrestrial multi-branch Tree of Life. It took 4 billion years. Microbes, plants, insects, fish, birds, animals, we all live thanks to the same macromolecules, the same materials of life: DNA, RNA, proteins, lipids, salts, and metals. Errors, failures, purges, and slaughters have occurred; but, once born, life has always managed to circumvent in some ways the traps of death that the environment sets out for it, to pass the survival tests to which natural selection continually submits it.

Chapter 3: The brief history of the human genome

Somewhere in these multi-millennial comings and goings, through the reciprocal influences between genomes and environments, among the species that roamed, appeared and disappeared in the backyard, modern man, *Homo sapiens*, emerged. It has been about 130-200 thousand years[25] that this big-headed, straight-spined beast perched on two legs, began its epic march and, after triumphing over its hominid roommates, has colonized the Earth. Yet he has existed only 1.4 hundred thousandths of the time of the Universe which envelops it, only 4.4 hundred thousandths of that of its host planet.

These are astronomical numbers! Surely, they are based on state-of-the-art observations, measurements, and scientific theories, with their own limits of certainty! They are numbers beyond our grasp nonetheless!

To gain some grip on these numbers, let us try an analogy: if the Universe had one week of existence – to borrow the time interval that, according to Judeo-Christian mythology, it took God to complete his work of creation – the Sun and Earth would have 2 days and 6 hours of age; man would have 8.6 seconds of life only. On the universal time scale, we are mere zygotes. In this scale, we are quite a recent event. What awaits us if we are allowed to grow to adulthood, in this scale? If, unlike so many other species, we survive the test of time?

Phylogeny is the filiation of species in the *Tree of Life.* Similarities and differences in the DNA sequences of homologous genes (specifying similar information) reflect this filiation. Closely related species share more similarities than distant ones. In the case of man, the chimpanzee is his closest relative. Both belong to the '*Homininian*' group, according to taxonomists. Their genes are 99% similar. Thus, it took merely a 1% difference between their genomes for them to form two distinct species. However, related as we are, the chimpanzee and we appeared in separate lines derived from a common primate ancestor, CHLCA (*Chimpanzee-Human Last Common Ancestor*), who lived 4 to 14 million years ago. The two lines had given rise to sub-lines. Ours, the *Homo* (Human) line, had also engendered *Homo habilis* 1 to 3 million years ago, *Homo erectus* 1 to 2 million years ago, and *Homo neanderthaliensis* as well as *Homo denisova* 100-200 thousand years ago. These last two *Homo* lines had been contemporaries of *Homo sapiens,* had lived and successfully mated with our line in Europe and Asia, but not in Africa. Indeed, Neanderthal and Denisovan DNA has been isolated from fossils and sequenced. Traces of it are found in modern-day Europeans and Asians, not in Africans. This finding only means that the two extinct lines roamed Eurasia, not Africa..

The point of this recapitulation is that, as *Homo sapiens*, we are a recent offshoot of a long, very long history of life. Indeed, since its appearance on the planet Earth, life, human life in particular, has carried chemical testimonies of its ambitions and wanderings in time and over territories: DNA and its variations.

3.2. These Africans that we all are

EVERYTHING THAT EXISTS LIVES, even the inert apparent. Life is motion, subatomic, atomic, molecular, cellular, or environmental motions. For unicellular or multicellular life, ecological contingencies – climate, tectonic, hydraulic and wind movements – motivate or contribute to these motions. Distances traveled vary by species' degree of autonomy and mobility. Those that fly, swim or walk are the most apt at it. Our two-legged ancestors, those brave Homininians or Hominins, took up the challenge with great determination: They traveled through forests, savannas, steppes, and tundra; they crossed streams, rivers, lakes, and oceans; they climbed mountains; they tumbled down canyons; they faced winds and gusts, rain and storms, heat and cold. They colonized the Earth.

Chapter 3: The brief history of the human genome

From the origins of life to the present day, the reasons for migrations have always been the same: the needs for survival. These needs can be summed up in three imperatives: the pursuit of prey, the escape from a predator, and the thirst for procreation. It is on these three points that the natural selection has always interrogated every species, in the past as well as today. The predator versus prey game is played out in the environment in its broadest sense: in the air, water, and land; among those who inhabit them and at their mutual expenses. For life usually feeds on life, by predation as well as by parasitism, commensalism or mutualism. The first two imperatives have the third for the final objective: to overcome annihilation by death, to project oneself in the future, to perpetuate oneself in the long term. It is by this supreme imperative that one measures the biological success of a species, because defeat to it would insure its disappearance pure and simple. In the course of eons, species that have failed this test have vanished.

However, from my point of view, and contrary to the widespread notion of the words, natural selection is not a struggle without respite, indiscriminate, and unforgiving. Instead, *it is a game of bioecological equilibriums that tend to favor the survival of most species, sometimes at the 'sacrifice' of a few individuals within each species.* Individuals can perish as long as the species persist. The continued existence of the predator depends on that of the prey. In the exercise of natural selection and competition among species, violence is often necessary, but it is never gratuitous. The survival of species, not of individuals, is the ultimate interest of natural selection.

Can the migrations of *Homo sapiens* since his emergence be traced back by examining his genome? The answer is yes: by determining the distribution of genetic variations, SNPs in particular, in different geographical territories. The premises of this assertion are: (i) *Homo sapiens* emerged in a given territory of the planet; (ii) he has lived there for several millennia, multiplying and accumulating a wide range of variations in his genome; (iii) waves of migration from the primary territory to a secondary one, from the secondary to a tertiary one, etc., occurring during limited periods of time each time displaced a successively smaller fraction of the original population and, consequently, a smaller fraction of the original number of genetic variations (Figure 3.2).

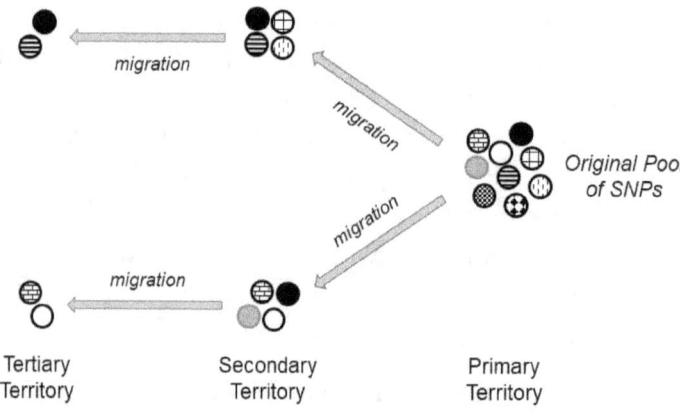

Figure 3.2. *Migrations and SNP partitions*

The genomes of the Aborigines from different territories of the world were of particular interest in this respect. The term 'Aborigines' comes from the Latin '*ab origines*' which translates as 'from the origins'. It identifies the first peoples established for millennia in a territory and preserved from genetic admixtures with other more distant peoples. Their genetic pool is representative of the ancient migrations that have distributed human SNPs. Scientists sequenced the genomes of these peoples and inventoried the SNPs they contained. Most SNPs were common to all peoples, many were infrequent in some, several were absent in others, but *all SNPs were present in sub-Saharan Africa*. By the logic of distribution of genetic variations through migratory waves, Africa would be the original territory of these migra-

Chapter 3: The brief history of the human genome

tions. According to genetics, therefore, Africa is the cradle of humanity. *We, humans, women and men, are all Africans.*

Other evidence amply supports this conclusion. The oldest confirmed fossils of *Homo sapiens* dating from 195,000 years ago were discovered in Ethiopia. In recent years, other older fossils of presumably modern man have been discovered at other sites, all African sites, e.g., in Morocco (Jebel Irhoud cave) and South Africa (Florisbad skull). In other parts of the world, they appear much later: 100,000 in the Middle East, 50,000 in East Asia, and 40,000 in Europe and 30,000 in the Americas. These are approximations, of course, but they are not speculations, based as they are on scientifically rigorous criteria such as the anatomy of fossils, their isotopic dating, the geological strata of preservation, as well as tools, animal remains, and artistic sketches or paintings found near them.

If this is the case, why then does the '*monocentrist*' thesis of the origin of humanity – otherwise known as the '*Out of Africa*' or OoA hypothesis – have skeptics and even opponents? On the one hand, there are '*pluricentrists*' who maintain that *Homo sapiens* emerged in multiple regions of the globe from homininian ancestors common to all territories or unique to each territory. They give as an example that the presence of the Neanderthal man's fossils in East Asia and Europe, of Denisovan man's fossils in South Asia, but not in Africa. On the other hand, there are proponents of the middle ground between 'monocentrist' and 'pluricentrist' theses. They contend that a 'pre-*Homo sapiens*' emerged in Africa, traveled to Asia and Europe where he met other hominins who had preceded him there, interbred with them, engendering, over the course of millennia, Asians and Europeans. Again, they cite as proof the recent sequencing of the fossilized genome of Neanderthal and Denisovan men, which has revealed that these hominins shares minute genomic fractions with Europeans and South Asians, respectively, but not with Africans.

We cannot reproduce the evolution of man in a small-scale experiment. Therefore any reasonable hypothesis on the subject deserves consideration, especially if offers novel avenues of investigation. However, the alternative hypotheses which question the OoA explanation violate a fundamental law of science: the law of simplicity. If all the accumulated evidence supports it, the simplest hypothesis is probably the right one. Parsimony, as I have already said, is a principle of Nature. There is no need to invoke convoluted models when a simple and straightforward model comes to mind. One wonders whether the biologists who dismiss the OoA model do so out of inner discomfort, conscious or unconscious, at the thought that they and all humans are Africans, out of a deep desire to affirm the original distinctiveness of one's own group, a case insistently argued by some prominent Chinese scientists. Yes, indeed, scientific investigations are often tainted with unspoken wishes and anticipatory desires impressed on minds by the ambient culture. Yes, sometimes it is with much reluctance or a broken heart that the most honest scientist bows to the weight of objective evidence.

3.3. Abebe and Abeba: the seminal couple

THE EVIDENCE IN FAVOR of the OoA explanation also comes from the geographical origin of the first man and woman. Since they appeared in Ethiopia, I will baptize them with the Ethiopian names of *Abebe* and *Abeba*, respectively. Both names mean 'flowers'. Metaphorically, they symbolize the original pollen and stigma, the first seeds that engendered the vast human garden. Scientists established their territory of origin by state-of-the-art genetic methods. Remember my earlier comments about chromosomes, about the Y chromosome that controls masculinity? It is carried only by men and transmitted only from father to son. Therefore, if one can find genetic variations on this chromosome, by applying

Chapter 3: The brief history of the human genome

the principle of 'the most diversity of the first-appeared' (as illustrated in figure 3. 2), one can trace the geographical territory inhabited by the first man. It was not easy. First, the Y chromosome is one of the four shortest (20, 21,22, Y), and half the size of the X chromosome. Second, of this quarto, it has the smallest number of genes, including SRY, which converts the fetus from female to male. Third, most of his DNA is made up of repetitive sequences, with little informational diversity, as if merely chaperoning a few isolated genes.

This genetic scarcity has long been a laughing matter for women scientists and feminists, who could not help but ridicule this pathetic little male organelle that has caused so much damage in human history. Here is a short story from personal experience. From November 1993 to October 1994, I was a sabbatical researcher in Dr. Elizabeth M. Simpson's laboratory at Jackson Laboratory in Bar Harbor, Maine. I went there to learn how to produce by genetic engineering a mouse lacking an enzyme responsible for male fertility. Simpson was one of the few scientists in the world trying to elucidate the little genetic information hidden in the Y chromosome. She had posted on her office door a poster of the chromosome. Alongside the image were names of putative genes. These included conceit, love of gadgets, sports, violent films, spiders, and snakes, reading in the toilets, selective inattention, emotional stinginess, and amnesia of dates: all subjects of amusement at the expense of the only male of the team that I was.

Getting back to the issue at hand, through tenacity, scientists eventually characterized about 50 genes on the Y chromosome (compared to about 800 in the X chromosome), inventorying a variety of SNPs in populations around the world. Once again, all the Y chromosome SNPs were present in Africa, while variable fractions of this African pool were scattered on the other continents. This survey of exclusively male SNPs indicated that *Abebe, the father of humanity, was indeed an African.*

THE COLORS OF GENES

What about Abeba? She too has left her mark in the genealogy of women. She had DNA that she passed on exclusively to her daughters and several times-great-grand-daughters through generations. You guessed it: mitochondrial DNA (mtDNA). It is short and circular like that of the bacterium; it contains only 37 genes which serve as blueprints for the manufacture of protein workers for mitochondria which, as discussed earlier, are furnaces that produce the chemical energy needed for cellular operations. These organelles come in multiple copies scattered throughout the cell cytoplasm. Men and women carry them; they inherit them from their mothers. Gametes – eggs and sperm – contain them too. When a man inseminates a woman, he pours into female genital conduits a crowd of tail-beating sperm which engage in a fierce marathon competition toward an expecting egg. When the victorious sperm meets the egg and penetrates it, it discharges into it only its genomic DNA contained in its big head (the acrosome), leaving all the rest out. The fertilized egg, the zygote born from this fertilization will therefore have only its mother's mitochondria as the energy source to feed its development. Thus, metaphor aside, mothers do carry and nourish life from the first instant of life and beyond.

The mitochondria we all carry come from a single female ancestor. Much like the Y chromosome, over millennia, this woman's mitochondrial genome had accumulated variations in its DNA sequence, creating a mosaic of variable mitochondrial genomes, some more populous than others. Scientists sequenced mtDNA of peoples from different geographical origins and inventoried their genetic variations. The mtDNA of Africans contains them all; that of more distant peoples, differing fractions of the African pool. Logic dictates: *Abeba, the mother of humanity, was an African.*

All right! Okay! Let's say Abebe and Abeba emerged in Africa. Let's accept that this continent was the first territory of

Chapter 3: The brief history of the human genome

human settlement on Earth. Is that important? Not at all. The first man and the first woman could have appeared anywhere else if conditions were favorable if alterations in a hominin genome under environmental pressures had made the jump to consciousness possible. They could have emerged in the Middle East, for example, in a mythical garden, the Garden of Eden, as would have it those who believe the Judeo-Christian biblical story; they could have been named Adam and Eve. That wouldn't matter. The central lesson of this research is that we are all the descendants of a seminal couple, that we are all brothers and sisters, despite our differences.

At first sight, we cannot help noticing that we do differ among us and in several ways: in skin color, hair texture, eye shape, nose aperture, lip thickness, etc. How did these differences appear in the human family? Geography, as we shall see, offers some answers. Are these differences skin-deep or deeper than they seem? Can they be used as criteria for a classification of the human species? Do they affect human nature? How does genetics respond to these questions?

3.4. The signatures of natural selection

THE ENVIRONMENT, as I have said, influences the genome and determines its expression from moment to moment, day to day, through the years, centuries and millennia. In the mid-term, it adjusts the level of gene expression through temporary genetic markings (epigenetic markings) to meet the individual's life needs. In the very long-term, it consolidates genetic variations that promote the survival and fertility of species. Natural selection, as Darwin had explained it, acts on this long term; it shapes and fixes the traits of species. However, other mechanisms may come into play: it sometimes happens that the geographical or cultural isolation (voluntary or imposed) of a restricted group within the species, leads to a narrow mixing of genes and a phenotype typical to this group. This process is called *genetic drift, bottleneck,* or *founding effect*.

These theoretical generalizations attempt to explain gene-environment interactions that occurred over a long past. They would be more convincing if they had the support of observational, experimental, and preferably genetic evidence. Of genetic evidence, there is plenty. Indeed, for a reasonably sized population in which genetic mixing occurs freely, natural selection has stamped its action on many genomic regions. In this age of easy and rapid genome sequencing, it has become possible to search for and detect such signatures. One of the most accessible of these signatures is called *selective sweeps*.

To understand this signature, let us take a brief step back. Let us remember that each one of us comes from the fusion of two half-genomes, our mother's and our father's. In all somatic cells, the half-genomes, subdivided into chromosomes, face each other in pairs of similar chromosomes (called homologs) and work together for the wellbeing of the body. In germ cells, however, this chromosomal face-to-face ends when they become gametes (eggs or sperm) during meiosis. Pairs of homologous chromosomes exchange segments, thus generating new pairs of homologous chromosomes inhabited by variable contents of paternal and maternal genes. These mosaic chromosomes make up the half-genomes of haploid gametes.

This mixing of the genetic material of paired chromosomes is called homologous recombination. It occurs more easily between distant regions along chromosomes, but little or not at all between adjacent ones. Thus, if in these distant regions, the paternal and maternal copies carry different genetic variations, these variations may end up on the same chromosome after meiosis. Homologous recombination is a random mixing process. Its absence would result in the presence of the same series

Chapter 3: The brief history of the human genome

of variations in chromosomal domains, generation after generation. This genetic conservation occurs when in regions containing genes that confer a remarkable advantage for survival and fertility. Natural selection tends to preserve them unaltered.

Figure 3.4 (below) illustrates this concept. In this example, let us assume, to begin with, that a gene is present in a population in three different copies carrying various neutral SNPs (filled circles), i.e., not conferring any survival advantage or disadvantage. Let us then suppose that, at a given point, a SNP which confers a remarkable advantage (empty circle) appears in one gene copy (step 1). Natural selection (NS), acting over generations, will favor recombination events that preserve this beneficial SNP. Through the process, it will increase its presence in the population, *taking along (sweeping) neighboring neutral SNPs* (step 2, boxed). Thus genomic sweeps are signatures of natural selection.

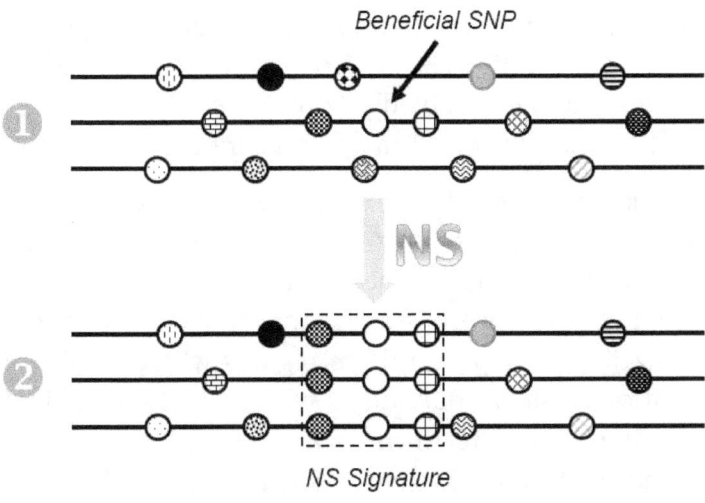

Figure 3.4. *Selective sweep*

Geneticists have identified a multitude of such sweeps in humans. Skin pigmentation is the most immediately visible feature of an individual. It is dark in populations living in tropical regions where the sun shines longer and is more intense; it is light in peoples living northern regions where the sun shines relatively less, with all imaginable epidermal gradients between these extremes. By pure empiricism, it was reasonable to think that pigmentation would be an adaptation to the climate. Are there any signatures of the natural selection of this trait? The answer is yes.[26] Melanin is a critical molecule in the pigmentation of the skin, hair, and eyes. It provides a protective screen against the noxious effects of solar ultraviolet (UV) rays on the skin. About forty genes participate in its production in specialized cells called melanocytes. These genes include *TYR*, *OCA2*, *SLC24A*; the first is involved in the manufacture of melanin; the other two, in its transport to the right place in the skin, hair, and eyes. Scientists have identified SNPs in these genes and revealed selective sweeps around them, indicating that natural selection preserved them. Notably, selection of SNPs of the *TYR* gene occurred in European and Asian populations (not in African ones, who didn't need it). In Eurasians, these SNPs and many others have undoubtedly contributed to the preservation and spread of skin pallor as an adaptive trait.

The skin manufactures vitamin D. The body needs it to strengthen its bones and for other functions. People lacking it are at risk of bone fragility, rickets, and osteoporosis. The skin manufactures it under the stimulation of solar UV rays. Melanin masks these rays. In northern regions where insolation and UV radiation are low, individuals with a highly melanized skin are more likely to suffer from vitamin D deficiency, a mortal risk in ancient times when an individual had only his legs to chase out his prey or flee away from his predator. There and then, skin pallor conferred a survival advantage and, as a result, was preserved in local populations. With large-scale migrations of modern times, highly melanized individuals living in northern countries show chronic vitamin D deficiency. Enriching their diets with the vitamin can correct the deficit.

Chapter 3: The brief history of the human genome

While solar rays are useful in the production of vitamin D, their overabundance is likely to cause skin cancer. Thus, in sunny regions of the globe, having a melanized skin offered a survival advantage since melanin is an effective barrier against UV rays. However, to be an adaptive trait, pigmentation had to have conferred greater childhood survival or adult reproductive success. Inversely, severe melano-deficiency should have caused higher childhood mortality or adult infertility. Skin pallor genes would have become scarcer in the population as time passed. Is it possible that, in sunny regions, skin cancer was a major cause of mortality of 'amelanic' children? Alternatively, could, in the distant past, 'amelanic' individuals have been excluded from contributing to the reproduction of the band or tribe by banishment or immolation?

Here a minor caveat is in order: when it comes to contemplating human evolution over 200-300 thousand years with the eyes of the present, nothing is so sure. We cannot recapitulate in a laboratory or a controlled environment the experience of this millennia-old odyssey. We can only analyze what today offers us, fossils and vestiges of the past, and try to connect them in a way that gives us a vision that best explains our emergence and our journey. The theories we elaborate and the concordances we establish serve to reassure us about the relevance of our vision. In other words, the selective sweeps of SNPs we have discovered around genes that regulate melanin production comfort our theory, but do not prove it. We could say the same thing of other trait differences we notice in human groups, be it the shape of our noses, eyes, and temples, or the texture of our hair, or the thickness of our lips. These traits are hereditary within each group and, to that extent, are undoubtedly determined by genes and the environment.

As scientists imagine plausible biological explanations for the origins of differences among human groups, they cannot resist the impulse to classify these groups, choosing extremes and ignoring gradations, arbitrarily and undiscriminatingly freezing them under the word 'race'.

3.5. The illusion of race

WHAT IS A 'RACE'? A population of similar epidermal pigmentation, with similar facial morphology and hair? Of common ancestral and territorial origin? Sharing a common cultural heritage? What is each 'race' made of? Do Jews form a 'race'? What of the Arabs, Ainus, Batwa or Hottentots? What about Latinos and Amerindians? How many 'races' does humanity count? Three, four, five, six? How many? Can science, as an objective referee, answer the question? Can genetics, the science of heredity, corroborate it?

It most certainly tried, with all the means at its disposal. And it concluded that, at least in the language of genes, *human 'races' do not exist*; 'race' is an illusion, if not an senseless error or a patent lie. Some would say: "Yes, but there are recognizable physical differences between human populations! Why can't these differences serve as classification criteria?" The answer is simple: these differences are so small, so tenuous, and so diffuse that they cannot form the basis for classifying human beings into 'races'.

Let us look at the genetic evidence. Let us go back to the variations, these small grains of variable colors scattered throughout our genome, which allow us to trace back our peregrinations on planet Earth. Let us accept for a moment that 'races' do exist and that they stem from genetic differentiation among them. In such a case, the variations found within 'races', collectively taken, should be notably different between 'races'. This is not the case. In fact, *there is more diversity within each presumed 'race' (85-99%) than between them (1-15%)*. This surprising

Chapter 3: The brief history of the human genome

observation reported in 1972 by the geneticist Richard Lewontin, then at the University of Chicago,[27] reduced the notion of 'race' to insignificance. Lewontin had studied the diversity of blood types, blood enzymes, and antibodies in populations of scattered origins. He concluded his report with these words:

> "The classification of human beings by race has no social value; it actively destroys social and human relations. Since such a classification has no genetic or taxonomic relevance, it should be abandoned.[28]

Subsequent surveys of genomic SNPs within and between populations of different ancestries have amply corroborated Lewontin's finding. Popular media relayed the finding with much fanfare; anti-racism militants quoted it *ad nauseam* claiming loudly and clearly: "Finally, science has proven that racial classifications have no biological basis! Shame to those who still refer to it! They are either ignorant, naive or ill-intentioned."

Was the debate extinguished after that? Far from it! Some have accused Lewontin of having tinted his interpretation of the data with his socialist – Marxist at some point – and egalitarian ideology. Others have timidly questioned his radical conclusion. But by far the most scathing criticism came 30 years later from Antony William Fairbanks Edwards, a geneticist at Cambridge University in England. I had the opportunity to see and listen to him during my sabbatical year at the Jackson Laboratory in 1994. This stern, authoritarian aristocrat did not beat around the bush: he dismissed Lewontin's analysis as superficial and treated his conclusion of erroneous and misleading.[29] To him, the survey of gene variations was not enough; it was essential to establish correlations among these variations. In doing so, according to Edwards, one can confidently classify individuals in distinct groups, which correspond rather well to current 'races'.

For the layperson, this debate between geniuses can be confusing. It is not uncommon in science. Indeed, the same data can lead to different conclusions depending on the sharpness and thickness of the magnifying glass chosen, on the assumptions made, and the statistical tools used. Quite often, this debate does not pertain to the accuracy of the data themselves, but to their interpretation, to the favorable emphasis placed on some data and dismissive treatment of divergent or contrary ones.

From my perspective, the human genome resembles our Earth: it is a rugged landscape, shaped by ravines, valleys, and canyons, mounds, hills, and mountains, crisscrossed by brooks, rivers, and oceans, but with no defined or unbridgeable boundaries. Genomic topology is in large part a reflection of territorial topology. It does not confine individuals of our species to classes and 'races', as it allows for genetic exchange and mixing among them. The same way Google's satellite camera can provide sharp pictures of Earth geography, modern DNA sequencing methods combined with efficient computer-assisted data processing can draw up a finely detailed genomic topology.

The genomic topography of an individual within a geographically, historically and culturally distinct group is called 'ancestry'. It reflects the various influences of the environment on the group's genetic pool, its pressures and selections. Genetic ancestry screening has become an industry. To curious dilettantes, it has revealed many surprises, unexpected admixtures that invalidate all traditional classifications into 'races'. This genetic admixture is all the more accentuated in intermediary territories (e.g., Mediterranean countries, East Africa and Middle East) or territories of recent colonization by peoples of different origins (e.g., North and South American countries). To the point that, in these territories, if one is not compelled to do so, one has to choose the 'race' under which one prefers to identify oneself, notwithstanding genetic proclamations to absurdity.[30]

Chapter 3: The brief history of the human genome

The former U. S. President Barak Obama is a typical example of a first-generation Caucasian/African American, who felt socially coerced and psychologically inclined to call himself solely African American. Another example is the ridiculous case of Mostafa Hefny, an Egyptian immigrant to the U.S., who, in spite of his black skin and curly hair was dubbed 'White' just because the U.S. law stipulates that people of North African origin are to be considered 'Whites'. Badly wanting to be reclassified as 'Black', Hefny, has sued the U.S. government for years in vain.[31]

The genomic topology of a population, I repeat, has been shaped over time by environmental demands to maximize the chances of survival. The hills and mountains of this topology (the most frequent genetic variations in a population) have served the needs of this population in this environment, even though in some cases and some forms these variations may cause individual morbidity. Moreover, when circumstances change, yesterday's strengths may become today's vulnerabilities. It is within such a framework that we must interpret the segregation of certain diseases in peoples of common ancestry. This segregation does not in any way disqualify the genomic pool of this population. Such a conclusion would be contrary to biology; it would be a perversion of genetics.

Chapter 4

The Perversion of Genetics

4.1. Eugenics gone amuck

As experimental geneticists, we remove genes from or add them to cells or organisms; we then examine the consequences of this manipulation on cellular and body functions. Our favorite experimental model organism is the mouse. This little animal recapitulates the biological functions and life trajectory of man fairly well, but in smaller dimensions and in less time: after 3 weeks of gestation, a mouse is born; 3 weeks later it reaches adolescence; it becomes sexually mature at 6 weeks of age, old at 12 months and elderly at 24 months.

A curious phenomenon has been observed when a female mouse becomes pregnant: for it to complete gestation, the male mouse that fertilized it needs to be kept in the same cage. Substitution of this mouse with another will compel the pregnant mouse to abort its fetuses (and sometimes devours them) and let the new male inseminate it three days later. The female 'interprets' the substitution as a victory of the second male over the first; its behavior is motivated by an instinctive desire to pass on the 'powerful' genes of the victorious mouse to its offspring.

This discriminatory behavior is an instinctive form of eugenics (from the Greek *'eu'* for good or well, and *'genos'* for birth or descent). Humans practice it, consciously or unconsciously, in person or through their families, when choosing a partner of the opposite sex with whom they hope to engender a healthy progeny that will perpetuate their genes. This perpetuation is more likely to succeed if the environment lends itself to it, if there is an abundance of conditions propitious to survival, e.g., access to food and social status; hence, the consideration given to these conditions in matrimonial enterprises in all societies.

Instinctive eugenics underwent a fateful deviation when human societies decided to use social status to influence the flow of genes, mostly to prevent their free flow: when it became institutional. The societies in question were the oligarchies of birth and power. *The goal of institutional eugenics was to make social privileges hereditary.* The effort goes back to the beginnings of history. Egyptian pharaohs married their sisters to preserve the purity of their blood. Greek Spartans used to get rid of children who were physically imperfect to ensure the martial vigor of their society. Indian Brahmins were to take spouses only within their caste of the 'privileged reincarnates'.

In the Western world, it is in Darwin's theory of evolution that eugenics found a scientific justification. In fact, it is Darwin's cousin, Francis Galton (1822-1911), who coined the word. A multitalented genius, he had developed statistical methods to assess the heritability of biological traits. Following his studies on twins and pedigrees,[32] he became convinced that talents and disabilities were hereditary. He proposed that, rather than waiting for the slow and late purifying effects of natural selection in favor of the ablest individuals, it was society's responsibility to accelerate the improvement of the human species (the 'British species', in particular) by encouraging reproduction among gifted people and discouraging that among ungifted ones.

Sociologists used the results of Galton's research to explain economic inequalities and justify social injustices, the submis-

Chapter 4: The perversion of genetics

sion of the female gender and the domination of foreign peoples. The results became a springboard for policies of large-scale ethnic cleansing implemented by some States during the last century. Euphemisms apart, the eugenic theory states:

> "Individual or collective supremacy, whether intellectual, economic, or military, is the result of natural selection acting on genes hidden in cells and transmitted through gametes. Social status is a product of heredity; it originates from genes; it is futile to attempt to modify it. Left to reproduce among themselves, aristocrats will engender aristocrats; commoners will engender commoners. When mixed with commoners' genes, aristocrats' genes invariably degenerate. Such mixing works against Nature, making improvement of the human species difficult. It is up to aristocrats to facilitate Nature's work by promoting their reproduction and slowing down that of commoners to the bare minimum needed for the labor force. Science teaches this; genetics commands it."

In this context, I remember a conversation I once had in my house with a young Mormon missionary. Very innocently, he tried to persuade me, an African, that it was God's will written in the Bible that Africans should forever serve their Caucasian brothers because they are descendants of one named Cham, Noah's cursed son who had laughed at his father's nudity. Nowadays '*Biologism*' is the new religion. Its omnipotent goddess is called 'Nature'. She distributes her blessings and curses as she pleases. Like Pontius Pilate to protesting Jews, she proclaims: '*Quod scripsi, scripsi.*'[33] Written where and how? Where else and how else? On chromosomal tablets in the AGCT alphabet of the genome, of course, the new bible. As in the old one, everyone reads in it what he wants to believe.

In French literature, the most charming troubadour of hierarchy in human nature is without contest Count Joseph Arthur de Gobineau (1816-1888), a French diplomat and writer. His opus entitled '*Essai sur l'inégalité des races humaines*',[34] is quite impressive by the vast panoply of examples, the vigor and limpidity of his argument. To illustrate the consequences of genetic pollution, he contrasts the underdevelopment of Latin America, where Caucasians interbred freely with Africans and Native Americans, compared to the development of North America where Caucasians discouraged such genetic admixtures by law, faith, and the whip.

The last great herald of eugenics is the American physicist William Shockley (1910-1989), inventor of the transistor, Father of the Electronic Age, and winner of the 1956 Nobel Prize for Physics. He used the fame conferred on him by the prize to cover of scientific authority his prejudices about individuals and groups. He sounded the alarm about the higher fecundity of 'imbeciles, white and black', which, in the long run, would reduce the intellectual capacities of America as a whole. He postulated that this degeneration was more accelerated in 'Blacks' because they are naturally less intelligent than 'Whites'. Denying that his views had any racist and malicious motivation, he wrote:

> "I sincerely and thoughtfully believe that my current attempts to demonstrate that American Negro shortcomings are preponderantly hereditary is the action most likely to reduce Negro agony in the future."[35]

He proposed that the intellectually disabled be sterilized. 'Generously', he suggested that this be done against payment: 1 000 USD for every 1% of IQ below the average of 100; he donated his sperm to an *ad hoc* bank as a contribution to the improvement of the human species. Unfortunately for him, he spoke out of time: the people of his time did not lend him a favorable ear; instead they listened to him and his theses with contempt and disgust.[36]

In this regard, I would like to grant the presumption of good faith to these aristocrats of condition and ideology, the Galton, the Gobineau, and the Shockley. They may not have been intentionally malicious in their thinking. Perhaps, from their stations in life, they observed peoples and societies; perhaps they honestly tried to understand and explain the differences among them to the best of their abilities, using (and sometimes unconsciously abusing) the scientific theories in vogue. Unfortunately for them, in science, contrary to popular saying, the exception always invalidates the rule; it opens a breach in the rule; it calls it into question. And exceptions to the classification and ranking of human groups abound; history, past and present, provides a plethora of them, as we shall see in the following sections.

Whatever the motives of the prophets of eugenics, the conversion of their theory into science-based ideology had desensitized the consciences of the well-meaning people in the face of the rejection, misery, hurt, and murder suffered by other human beings, degrading them to the status of circus beasts. Emblematic of this degradation is the case of the South African Khoikhoi woman named Sarah Baartman (and disparagingly nicknamed 'Hottentot Venus', c.1779-1815) who, because of her remarkable steatopygia (excessive fat accumulation in the posterior), was displayed as a freak for 5 years in England, Ireland and France. So is the case of the Congolese short-statured Mbuti man named Ota Benga (c.1883-1916), brought to the U.S. to be put on display as an anthropological curiosity at an exhibit in St. Louis, Missouri, and later at the Bronx Zoo alongside an orangutan.[37]

Eugenics had nourished the conquering ambition of peoples and countries. It had legitimized the cruelties of slavery, the violence of colonization, the brutal appropriation of the world's resources, the bloody genocide of the Hereros and Namas of Southern Africa (1904-1907), long before the more 'sanitized' one of the Jews of Europe (1939-1945). The jungle 'rule of the strongest', which had too often governed relations between individuals or groups, was turned into a predestined noble mission commanded and authorized by genes, those good genes that must be preserved by the 'breeding' of pure-blooded progenies. Ancient Rome of Marcus Aurelius (121-180) had contemplated such 'racial hygiene'. Adolf Hitler's (1889-1945) Germany secretly attempted it through the *Lebens Born* (Fountain of Life) Project with its 'golden farms' where Nazi officers fertilized Aryan women to produce superior children, future citizens of the *'Deutschland über alles'* [Germany above all]. This litany of the depravations of eugenics is more than an accusation or condemnation; it is a testimony to the 'ineffable' ability' of man to rationalize his most selfish and cruel choices.

In its less aggressive form, eugenics has used the words and tools of genetics to classify types and groups within the human species and to quantify their 'humanity' by measuring and comparing their physical and mental abilities. Over the last half-century, comments and debates on this subject have crystallized on intellectual differences between Africans and Europeans.

4.2. Obsessing about IQ

BESIDES CONSTANT WALK on two feet (bipedalism), the volume of man's skull is the other physical feature that distinguishes him from his cousin the chimpanzee. It has three times the capacity of that of the ape (on average 1400 cm^3 (range: 1100 to 1700) compared to 387 cm^3 (range: 325 to 500). It contains a larger and more sinuous brain. Psychologists and anthropologists have attributed to this organ performances that are believed to be exclusively human: the ability to think and reflect, to conceive and construct sophisticated tools, to anticipate and envision the future. They assumed that there was a direct correlation between skull capacity and intellectual ability. They measured and compared the skull capacity of different humans and, from these measurements, inferred their mental aptitudes.

For obvious reasons, this craniometrics preferentially targeted women and Africans, comparing them to Caucasian men. The measurements indicated that women had a 10% lower, and Africans 6% lower cranial capacity than Caucasian men. Having a smaller skull and therefore a smaller brain, women were judged to be intellectually inferior to men. Of course, society knows better nowadays since mounting social evidence has amply contradicted this assessment, and societal taboos imposed by the rise of militant feminism has forbidden it. As far as the intellectual ability of Africans is concerned, despite the ambient decency of political correctness which keeps many believers in its inferiority from affirming it aloud, it is for some self-assured intellectuals a legitimate subject of 'scientific' investigation, partly in view the socioeconomic as well as technological standing of Africa and Africans in today's world. To put it bluntly, they ask: could this precarious standing be due to the intellectual inferiority of people of African descent?

In this investigation, *'Race, evolution, and behavior: A life history perspective'*,[38] the book by Canadian psychologist J. Philippe Rushton (1943–2012) of the University of Western Ontario, has become an obligatory reading for anyone preoccupied by the ideology of human classification. Rushton compiled an exhaustive list of physical, intellectual and social differences among the three main ancestral groups which he called Caucasoid, Mongoloid, and Negroid, and which, in this book, I will identify according to their original ancestral territories as Europeans, Asians, and Africans, respectively.

The author acknowledges that these differences concern averages, that for each parameter examined in a given ancestral group, there is a quantitative distribution in the form of a bell curve as illustrated in Figure 4.2.; and that the bell curves of different groups overlap, but only partially.

THE COLORS OF GENES

Inspired by the much celebrated or vilified book entitled *'The Bell Curve: Intelligence and Class Structure in American Life'* by Harvard University psychologist and sociologist **Richard J. Herrnstein** (1930-1994) and American Enterprise Institute political scientist **Charles A. Murray**,[39] Rushton proclaims that the 'intelligence bell curve' of Africans lies to the left (towards lower values) of that of Europeans or Asians. Furthermore, he comments, the curves for various other parameters have revealed that, compared to Europeans and Asians, Africans reach physical maturity earlier and are more corpulent on one hand; on the other hand, they are more prone to crime, promiscuity and social disorganization, have a smaller skull and are therefore less intelligent.

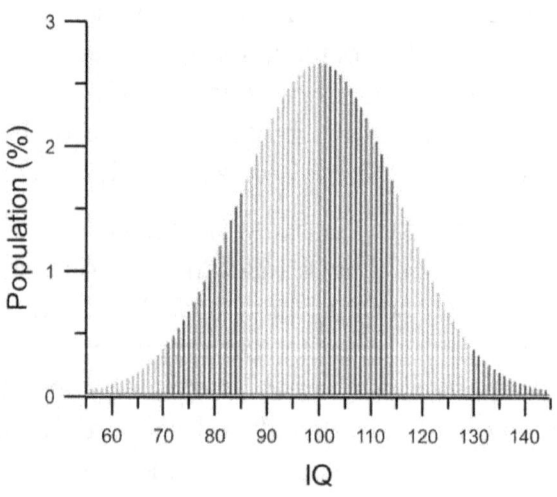

Figure 4.2. *The IQ Bell Curve*

In Rushton's own words:

"The reason why Whites and Orientals have larger hips than Blacks and therefore are less successful runners is that they give birth to babies with larger brains. During evolution, an increase

Chapter 4: The perversion of genetics

in cranial dimensions required women to have larger pelvis widths. What's more, the hormones that give Blacks an edge in sports are the same hormones that make them restless at school and prone to criminality."[40]

According to Rushton, the benefits of a bigger brain has been more pronounced in Asians than in Europeans. He argues that during the evolution of *Homo sapiens*, Eurasia's cold and inhospitable climate of Northern territories has fostered the emergence of a brain that is better equipped to guarantee one's survival and reproduction. The gift came at the cost of a relative reduction in sexual vigor, physical maturation, and aggression. In contrast, blessed with a gentler climate and a nourishing environment in the prehistoric tropics, Africans did not need the favors of such an efficient brain and therefore did not develop it. Rushton's acceptance of environmental influence is not in any way a negation on his part of the role of genes as causative agents of these differences. Far from it. In fact, according to him, many of these distinctive traits are passed on down generations: they are definitely hereditary.

By the way, it is noteworthy that funding for Professor Rushton's research originated from the Pioneer Fund of Wickliffe Preston Draper (1891–1971), a textile tycoon and an activist devoted to causes of eugenics, racial segregation, and selective immigration. Follow the money', goes the saying. Yes, it is sometimes difficult to separate scientific pursuits from their funding sources. Indeed, trends, ideologies, dominant theories, and money powers can contaminate scientific investigations and their results. Moreover, by alliance or complicity, agencies more easily finance investigations that meet their ideological expectations.

The scene was set up, and the word pronounced: genetics. All we had to do was wait for some smart biologists to identify the gene or genes for physical maturation, sexual vigor, aggression and, even better, intelligence. Some of them are already feverishly working on it, taking advantage of the new computerized tools of global analysis in molecular biology, the so-called 'omics' – genomics, transcriptomics, proteomics or metabolomics – which allow us to examine all genes and gene products at the same time, and derive correlations among them. Soon a few smart scientists may present us with a genetic perspective on our intellectual potentialities. However, the task has proven insurmountable so far. One reason may be that scientists keep asking the wrong questions. They remain stuck on the idea that there is a gene for each biological trait or function when evidence shows that the genome does not contain enough genes to satisfy such a scheme. Indeed, it is likely that what makes us human is expressed through the incommensurable interactions of the multidimensional structures of gene products.

While 'gene hunters' forage the genome for the 'Holy Grail', psychologists claim to have been able to measure the intellectual abilities of individuals, to have designed intelligence quotient (IQ) tests that they are both objective and unbiased. In scholarly circles, the uses and relevance of these tests have been the subjects of heated debates that are not about to subside. For most laypersons, it is an acceptable measure of the cognitive capacity of individuals, of their 'innate' intelligence.

In simple terms, mine, the intellect is 'a thinking machine'. Intelligence is the functioning of the intellect. The products of the intellect are thoughts, abstract and theoretical, useful and practical, immediate or anticipatory thoughts. Thoughts are the expression of intelligence in a living environment. As with all mechanisms selected by evolution, immediate and future survival is the ultimate goals of the 'machine'.

Where is the intellect? And in which organ? In the cranial box surely, the seat of the brain, that fatty mass crisscrossed by hundreds of billions of neurons variably interconnected in a thousand billion ways, linked by nerves and chemical messen-

Chapter 4: The perversion of genetics

gers (neurotransmitters) to all body parts, controlling and responding to them.

But which brain? Indeed, the brain is triune: it is reptilian, limbic, and cognitive. The reptilian brain (the basal ganglion) controls our instincts and reflexes of prey or predator. The limbic brain (which includes, among other things, the hypothalamus, the amygdala, and the hippocampus) directs our autonomous functions such as breathing and heartbeat, as well as emotions associated with nutrition, reproduction, and parenting. Finally, the cognitive brain (the neocortex) constitutes the real 'thinking machine'. It is this last brain which is most developed in *Homo sapiens,* it is it that had allowed him to invade, conquer and colonize the planet and its surroundings with more success than any other species.

Can intelligence be reliably measured? Yes, some people would say. Perhaps, would venture others, but only one form or another of intelligence, but never Intelligence, which remains an undefined notion. No, would proclaim still others, because, in its functioning, it is as moving as the environment – material, psychological, and psychic – that envelops it. Moreover, would add the 'negationists', one must keep in mind that the effort of measurement will always be tainted by the viewpoint of the measurer, by his culture, his life experience, and his expectations, even if he defends himself of undue bias.

If despite these reservations, one accepts the validity of the measure, one must also agree with the results drawn from it. These results indicate that the average IQ score of Africans is consistently 11 to 16% lower than that of Europeans. Is this difference explained by genetic or environmental differences between the two ancestral groups? There lies the problem: the choice of either one of these two options is necessarily a bias, and the conclusion, a prejudice. For, in either case, the supposedly scientific statements of psychologists and psychiatrists are mere hypotheses with no biological verification in view. '*Heredi-*

tarians' cite data that demonstrate the intergenerational consistency of the IQ differential between 'races'. Using their own data, *'environmentalists'* certify that this difference diminishes from year to year, as living conditions level out; and that monozygotic twins, separated at birth and reared in different situations, demonstrate it too.

If one rejects the validity of the measurement and its outcomes, any discussion becomes an exercise in pointless semantics. One could refuse to take part in it, to ignore the subject altogether, leaving the entire forum to *'hereditarians'* whose discourse might convince uninformed minds of its alleged veracity. I have chosen a better course in this essay: I will attempt to refute the validity of the test and the results.

However, in spite of all efforts, the controversy will not die out. It will often resurface, sometimes with unsuspected flagbearers, from some illustrious scientists such as Dr. James Watson who, in an interview with London's Sunday Times newspaper in October 2007, stated:

> "I am inherently gloomy about the prospect of Africa because all our social policies are based on the fact that their intelligence is the same as ours, whereas all the testing says not really."[41]

Unrepentant, he reaffirmed his views in a PBS documentary aired in January 2019, stating:

> "I would like for them to have changed, that there be new knowledge that says that your nurture is much more important than nature. But I haven't seen any knowledge. And there's a difference on the average between blacks and whites on IQ tests. I would say the difference is, it's genetic."[42]

Watson isn't just anyone: in 1953, with Francis Crick, he had elucidated the structure of DNA which ushered today era of molecular genetics. For many people, and me too, these statements were not simple slips of the tongue. They reflected the man's profound convictions based, not on scientific facts, but on his social itinerary as a petty bourgeois intellectual, draping his prejudices with a pretense of scientific detachment and objectivity. The world shouted shame on these words that many believed of another age, at least for enlightened minds. The comments tarnished the halo of the Nobel Prize winner. They initially cost him his job as director of the Cold Spring Harbor Laboratory and, lately, several honorary titles bestowed on him by the Laboratory.

4.3. But is it true?

IN TRUTH, RESEARCH ON IQ is a tendentious and pernicious extrapolation of Darwin's theory of evolution. It seeks to explain social inequalities by the law of survival of the fittest. In short, it proclaims the following:

- Groups which occupy subordinate statuses in societies are there because they are less adapted than the dominant groups to survive in the modern world.
- Because *Homo sapiens* triumph comes from the intelligence embodied in his brain, subordinate groups must be intellectually handicapped compared to dominant ones.
- Because the subordination of these groups persists over generations, the poor quality of their genes must be the cause; it is hereditary because it is genetic.

We may know nothing about the genes involved; we may understand nothing about the interactions between genes and the environment. We may grasp nothing about the intermolecular dialogues between gene products; as long as it suits our ideology, we go with the theory (in this case, natural selection) as the only frame of thought and vision. And, in so doing, one builds himself visors that prevent one from seeing anything else. Facts, in this case, do not guide the theory, but the theory guides the facts or at least their perception. This is an inversion of the scientific process; it is a perversion of science by ideology.

The debate on the genetic disparity of intellectual abilities among human groups mostly focused on Africans, in comparison with those of peoples with a less 'melanized' epidermis. One can assume that this skin difference facilitated and justified the comparison. I read the abbreviated form of Professor Rushton's book. It gathers old and current data in favor of the 'IQist' thesis and lists the 'causes and effects' of the intellectual differences between Africans and other 'races'. Having read it on an e-reader in e-book version, I can identify the chapters, but not the pages, from which I have fetched the excerpts quoted in this essay. I will quote the salient points of his discourse. To each, I shall oppose my arguments.

A. Rushton writes:

> "The early explorers in East Africa (the Arabs) wrote that they were shocked by the nudity, paganism, cannibalism, and poverty of the natives. (..) Several hundred years later, European explorers (..) wrote that Africans seemed to have a very low intelligence and a few words to express complex thoughts, (had) no wheels for making pot, grinding corn, or for transport, no farm animals, no writing, no money, no numbering systems. (..) The poor conditions of African countries and Black America have become a concern to many. (..) Neglect and decay are seen everywhere in Africa and much of the West Indies. (...) In the age of computers, fax machines and the world wide web, getting a dial tone in African cities is difficult."[43]

To that, I answer:

In fact, Arabs and Europeans rediscovered sub-Saharan Africa between the 12th and 15th centuries at the nadir of its evolution. They met her in her moments of high vulnerability. They approached her to trade with her in goods and people; then they sabotaged her, emptied her or subjugated her by slavery and colonization. It is gross ignorance or pure malice to

Chapter 4: The perversion of genetics

judge Africans in the course of human history by their situation over the last five centuries. Any student of human history knows that rise and fall are the fate of all civilizations, that the supremacy of societies changes. These historical vicissitudes derive from the environment, circumstances, or humans. Surely, Africa, as a continent, had known its periods of glory. For the lesson, consider the cosmic leap to consciousness experienced 200 millennia ago in East Africa by two named Abebe and Abeba, whose progeny mastered fire and, by inventing increasingly useful tools, triggered the most fantastic epics of life on Earth. Remember the tri-millennial grandeur of ancient Egypt (from 2300 to 30 B. C.) with its majestic pyramids and religious intuitions which inspired Judaism, and by filiation, Christianity and Islam, the two sprawling and proselytizing religions of today.

In this context, I must confess that, for too long and out of pride, I was, like many other Africans, reluctant to consider Egyptian civilization as African, having conceded its history to the Eurasian majority of its present population. About Africa's place in history, I wrote in my essay entitled '*Signposts of Hopes and Illusions*':[44]

> Modern ethnic Africa claims its kinship with ancient Egypt; but modern Egypt denies or ignores this filiation, considering itself an enclave out of the continental space. Does this paradoxical situation of a people claiming the heritage of other people has past precedent in historiography? With all due respect the great Sheikh, this tireless excavation for cultural continuity between Egypt and sub-Saharan Africa resembles a pursuit of glory by proximity. Between neighboring cultures, there have always been meetings and sharing, lending and borrowing. However, each culture has a heart and a soul of its own that it can claim without undue insistence. Africa possesses its Sundiata Keita, Nzinga Mbandi, Monomotapa, and Shaka Zulu to fill its history; it does not need the likes of Akhenaton, Tutankhamen, Charlemagne or Napoleon.

I have changed my mind on this point. I read more carefully the works of the Senegalese scholar Sheikh Anta Diop (CAD) (1923–1986) on the cultural continuity of Africa from ancient to modern times. Like him,[45] I denounce the unrelenting falsification of this continuity, the painting of Egypt as 'an accident of geography' with no cultural connection to the rest of the continent. Nevertheless, for reasons expressed earlier, I remain reserved about the obsessive insistence of the Senegalese scholar on classifying human groups according to their epidermal colors rather than their prehistoric territorial origins.

Let us have a 10-century panoramic view of the history of Africa. In West Africa, think of the territorial expansions under Sundiata Keita (1190–1255), the splendors of the Malian empire under Mansa Moussa, (c. 1280–c. 1337), the heroic resistance of Samory Touré (1830–1900), and yes, Timbuctoo, the pearl of the desert, the vibrant medieval city of commerce, knowledge and reflection until the 16th century. In Central Africa, contemplate for a moment the administrative genius of the Mani Kongo, Mwat Yav, and Balopwe of the Ne Kongo, Kalunda, and Baluba kingdoms, respectively (1400–1889). In Southern Africa, the mysterious Monomotapa of Zambezi may one day reveal the secrets of their walled cities. If the reader cares to explore further the contents of this short enumeration, he will note that in spite of the current challenges facing them, Africans have a long and continuous history, memorable and admirable in many respects.

Some observers of Africa suffer from having to concede to the African genius any creative work found on the continent. Casually, they attribute it to some imaginary foreigners, terrestrial or extraterrestrial. However, here, as elsewhere, the simplest explanation is the most plausible: any work of man, on or under a soil, comes from the inhabitants of the soil.

Like any other human society, Africans have tried to explain their place in the Universe and to design a destiny for

Chapter 4: The perversion of genetics

their eternity. The Dogon cosmogony inspired by their mastery of astronomy, the Baluba cosmogony with *Maweja*, their personal God and *Mikombo wa Kalowa*, the Redeemer, who was consumed to reconcile God and man,[46] Bantu philosophy of organic vitalism[47], ancient proverbs of Africa: these concepts are fascinating by their depth and wisdom, provided one takes the time to study them.

As far as the lack of connectivity with Africa is concerned, time since the writing of Rushton's remarks has overwhelmingly contradicted him. The continent has absorbed to its advantage modern technologies of information and communication and is connected to itself and the world, each day more and more. Which goes to show that a cliché of the moment is neither a reflection of the past nor a mirror of the future of a human society.

B. Rushton adds:

"INTERPOL Yearbooks show the rate of violent crime (murder, rape, and serious assault) is four times lower in Asian and Pacific Rim countries than in African and Caribbean countries. Whites in the United States and in European countries are intermediate."[48]

To this, I counter:

I cannot judge the assertion without examining how the author compiled the data. I assume that this 'prize' for violence and crime goes to the countries cited for personal offenses and occasional atrocities. It is unclear whether or not these crimes had for motivations legitimate revolts against established injustices or the needs of mere survival. Strangely enough, the history of the West (which I know better than that of Asia) has always seemed to me like an unending series of warfare that culminated in the slaughters of the two world wars. It is as if violence by individuals was conveniently illuminated by the lantern of the law of the dominant group, while that violence by the dominant group is covered up with sanctimonious justifications, with vilification of the opponent or the victim. As a result, mass crimes take on the appearance of either noble missions or anecdotal events. Examples include the Inquisition, the Crusades, the Slave Trade, the cleansing of the Americas, the massacres and serial amputations of Congolese populations under the Belgian king Leopold II, the genocide of Namibians under the Second Reich, the Shoah, the atomic bomb, etc. Already, I can safely say without risk of deluding myself that, when it comes to mass crimes, the prize certainly does not go to Africa or Africans. Too often, violence planned and executed by an organized group (Church or State, or both) becomes a crime only when the group is defeated.

C. Rushton continued:

> "IQ tests measure intelligence and predict real life success. The races differ in brain size and IQ tests. (..) IQ tests are made to have an average of 100. The "normal" range goes from "dull" (IQ around 85) to "bright" (IQ around 115). IQs of 70 suggest handicap sign, while QIs of 130 and above predict giftedness. The average Oriental IQ is about 106, the White IQ about 100, and the Black IQ about 85. (..) While Orientals developed complex societies in Asia, and Whites produced complex civilizations in Europe, Blacks Africans did not. (...) The average IQ of 70 for Blacks living in Africa is the lowest ever recorded."[49]

To this, I oppose:

Thus, according to Rushton, Africans would be mentally handicapped! Really? The work he cites is by the British psychologist Richard Lynn, 'a bird of a same feather'. Lynn collected IQ scores from around the world over a 20-year period. Oddly enough, Rushton forgets to mention how universally discredited this work has been concerning the truthfulness of its content and the quality of its statistical analyses. Also oddly, when Rushton correlates brain volume with intelligence, he refrains from referring to the lesser volume of women's brains

Chapter 4: The perversion of genetics

and the alleged lesser intelligence of this gender claimed by Lynn, and which current realities have convincingly contradicted. No doubt that he was intimidated by the current trend of female affirmation. He multiplies citations on brain measurements: cranial volume and circumference, cadavers' brain weight, magnetic resonance imaging (MRI), number of circumvolutions and surface area. He notes the concordance of differences among 'races' in all these measurements. By extrapolation, he even derives 'interracial' differences in the number of neurons in the brain: he claims that, on average, 'Blacks', having a smaller brain, would have 400 million fewer neurons than 'Whites'! From these physical differences, in a most strange quantum leap, he concludes that intelligence differs among 'races' as demonstrated by IQ tests.

To his credit, Rushton is unabatedly consistent in his biased inferences. He fervently recites articles that support his thesis and ignores with supreme indifference the vast literature that contradicts his claims. He blames it on liberalism and political correctness. For my part, I prefer to attribute the psychologist's rhetoric to an obsession with correlations. *However, correlation is not causality, nor even association.* For example, on the Belgian Congo of old, the prevalence of diabetes used to correlate with economic standing. Thence, the once widespread joke that money gave diabetes. We now know that several modifiable factors that would break the correlation between the two conditions, the most common ones being nutrition and physical activity. What do we know of the number of neurons, their connections, and plasticity? So little! The elephant or dolphin shows a higher ratio of brain to body weight than humans, what can we conclude about their intelligence? Nothing! What are we talking about when we talk about the environment: climate, nutrition, social stress in all its forms, beliefs, and existential priorities? Can we draw bell curves of 'races' that take all these variables into account? I doubt it! If even we could, would these curves be superimposable on those of IQs? And in what directions and ranges? We don't know! Will we ever know? Well, maybe. But, for now, the question remains. Anything said would be empty and unproductive speculation.

I can only guess what Rushton, had he lived, would have made of the recent discovery that populations of European and Asian descents carry short pieces of the Neanderthal genome while those of African descent do not, and of the fact that some genetic variations in these DNA pieces seem to correlate with anatomic brain structures involved in cognition.[50] I suspect that he could have branded the findings as 'additional scientific evidence' of the gift of a better intelligence granted by evolution to Europeans and Asians, notwithstanding the fact that previously, and for a very long time, the term 'Neanderthal' has been associated with coarseness and stupidity

Theorizing is the favorite sport of academics. They indulge in it to advance science. However, scientific rigor requires that an academic refrain from speculations founded on partial and blurry data, that he or she be honest enough to recognize, if not provisional ignorance, at least differences in data interpretation. Could psychology and psychiatry be sciences only in the impure sense of the word? Could they be mere exercises of intellection of observed behaviors and skills of human beings, under the guise of a biological and mathematical terminology? Their theoretical ramblings recorded in treatises over the years suggest that these disciplines, with their many turnarounds and reversals, are still guided by preconceived ideologies and are far from becoming objective sciences. The most striking reversal is the recent conversion of homosexuality from a 'disease' to a 'genetic predisposition'!

D. Rushton explains:

"Heritabilities, cross-race adoptions, genetic weights and regression-to-the-average all tell the same story (.. i.e.) that the genes cause race differences in IQ. (...) (Studies of) identical twins who

Chapter 4: The perversion of genetics

grow up in different homes (have shown that) IQ was 70% heredity and 30% environment.(...) One well-known trans-racial adoption study is the *Sandra Scarr's Minnesota Project*. The adopted children were either White, Black, or Mixed-race (Black-White) babies. The children IQ tests when they were seven years old and again when they were 17. (..) At age of 7, their IQ (of Black children) well above the Black average of 85 and almost equal to the White average of 100. (..). Black children raised in good homes had an average IQ of 97, but Mixed-Race children averaged an IQ of 109, and White children an IQ of 112. (..) At age 17, adopted White children had an IQ of about 106, Mixed-Race adoptee an IQ of about 99, and adopted Blacks had an average IQ of 89".[51]

To that, I reply:

It is possible that the trait measured during these studies (and I reserve the right to deny that it was intelligence) may contain a hereditary component. There are indeed physical matches between monozygotic twins in the anatomical structures of the brain, as measured by MRI, but not in their IQ.[52]

In any case, hereditary is not synonymous with genetic and vice versa; one is only a possibility of the other. In previous chapters, I have mentioned epigenetic factors that may influence gene expression during an individual's lifetime. These factors include, among others, the economic, social and psychological context of the mother during gestation, the family environment during the first few months of life, and the trauma associated with the adoption process. The studies cited did not consider these parameters. How could they? The parameters are too numerous and most of them not measurable. Instead, the authors have confined themselves to amalgams and averages, from which they drew somewhat dubious conclusions.

According to the Scarr study, this measure of intelligence of 'Black' children improved somewhat when these children grew up in 'good average White families' (cultural admixture) or when they carried the 'good white blood' (biological admixture). By symmetry, the IQ measures in cases of reverse adoption were necessary. What would happen to the IQ of disadvantaged 'White' children placed in 'good average Black families'? Such cases are rare, if nonexistent and socially unacceptable.

Note that, for the sake of argument, the investigators in this study chose to create a mixed-race category. They pretended to ignore the infamous 'One-Drop Rule', a social and legal principle according to which anyone who has a 'negro' ascendant is a 'Negro'. This eugenic principle, called *hypodescendance*, has been extended to any coupling between 'the superior White race' and any other 'inferior race'. According to this inexorable rule, Mulattos were 'Blacks' in the American society. I presume that, in this field of inquiry, sociological categorization can be revamped to fit one's preconceptions, contaminating the research and invalidating its results.

Rushton also cites the *'regression of averages'* to support the thesis of IQ heritability. To grasp this concept, consider the height of children relative to that of their parents. Two parents who are taller than average for their group will give children who are taller than that average, but shorter than their parents. Two parents who are shorter than average for their group will give children who are shorter than that average, but taller than their parents. This regression is due to the mixing of parental genes and the allocation of different combinations among children. In other words, the height of children in a given reproductive group will always tend to approximate the average size of the group.

Applied to the so-called IQ, Africans in general aggregate around their average IQ of 85, Europeans around theirs of 100, and Asians around theirs of 106. However, height is a linear, continuous and quantifiable physical parameter, while IQ is a diffuse mental parameter, tainted with subjectivity in both its definition and measurement. Moreover, for both height and the so-called IQ, this prediction becomes inappli-

Chapter 4: The perversion of genetics

cable when the group its limits are indefinable and environment is factored in.

To illustrate the importance of environmental factors, between 1930 and 1980, the average adult height, which is a recognized hereditary trait, increased by 4.1% in rich countries, by 1.2 to 2.4% in emerging countries and not at all in developing countries.[53] Why couldn't IQs be influenced by the environment too?

Finally, Rushton rationalizes 'racial" IQ differences by invoking '*r-K strategy*' theory, formulated by Edward O. Wilson, the famous Harvard University sociobiologist. In short, the theory states that the reproductive success of a species depends on a unique balance between its fertility ('r') on the one hand, and its capacity to take care of its offspring ('K') on the other hand. This balance may tilt to one side or the other. Sexual instinct governs the 'r' strategy, and parental instinct, the 'K' strategy. Because parental care is demanding, species with a 'K'-bent strategy have developed a more complex neural system and a more efficient brain than those with 'r'-bent one.

Rushton blithely seized this theory and frivolously applied it to the human 'races'. He concluded that, in Africans, the strategic balance is leaning more towards the 'r' type, whereas in Europeans and especially in Asians, it is leaning more towards the 'K' type. That is why, according to him, African children develop more rapidly, reach sexual maturity earlier, why African boys and men have a higher level of blood testosterone, are more muscular and more prone to violence, have more developed primary and secondary sexual organs, show more unbridled sexuality, reproduce more disorderly, and neglect more their offspring. Since their original habitat and way of life had not 'demanded' it of their genome during evolution, they did not acquire a brain as intelligent as that of the Europeans and Asians. All this, on average, of course, says the psychologist defensively, with no preference nor prejudice as far as the intelligence of any one individual is concerned. "All ancestries show favorable or unfavorable deviations from the averages, but there is no denying the deviations is skewed towards the unfavorable for Africans", he claims.

While we are at it, why not do it? So goes the saying. What acrobatic associations! What a biased collection for the sake of an ideology! Unbridled sexuality, disorderly reproduction, offspring negligence? At times, one has the impression of reading a monograph of an anthropologist's encounter with the 'savages' of a world he has just discovered, whom he evaluates and judges by the standards of his own culture. This reminds me of a Belgian ethnologist, a missionary priest by vocation, who had concluded that Baluba people were emotionally deficient (neither love nor compassion) because, according to his research, they had only one word (*kuswa*) and his negation to express affection and its absence.[54] As a true heir of the language, I have identified several words of variably nuanced meanings that express feelings [*ku* (to): *ambisha* (court) *anisha* (appreciate), *bedia* (disdain), *kama* (admire), *kina* (hate), *kisa* (torture), *nanga* (love), *nyoka* (detest), *samba* (console), *tenda* (praise), etc.}.

Knowing the African culture as I do and having been brought up within it, I fail to recognize myself in any aspect of Rushton's denigrating tirade, even if he drapes it in scientific terminology. This is not bad science; it is not science at all, period. Therefore, I will not honor the psychologist's reckless insolence and foolhardy extrapolations with an elaborate argument. Their obvious ridiculousness constitutes a better response than any contradicting discourse. Instead, I choose to dismantle the terminological foundations of his ideological '*hereditarism*' by clearly establishing what genetics says and doesn't say.

Chapter 4: The perversion of genetics

4.4. The assertions of genetics

MODERN GENETICS is still in its infancy, but already its current exploits bode for an interesting future in spite of the dangers of misuse of its findings. Today, for a modest price, it is possible to determine the 3 billion nucleotides that make up a person's genome; it is possible to estimate his or her real biological age by measuring the length of his or her telomeres (typical DNA which is located at extremities of chromosomes and get shorter with the passage of time); it is possible to determine his or her ancestry or ancestries by detailed mapping of his or her genetic variations. More and more people engage in this genetic prospecting for the pleasure of it. Some among them look for much more: they want to browse over their long-term genetic horoscopes, to consult DNA oracles on their biological blessings or curses: the promises of health and longevity or the threats of disease and early demise. However, like horoscopes and oracles, genes speak in sibylline sentences that one must interpret with circumspection and discernment.

What about genetic diseases? First of all, let's say that severe genetic diseases are relatively rare. As a general rule, life is granted to viable bodies only. To severely defective bodies, it is denied very early on, at the stage of the zygote, the embryo, or the fetus. This is why each living person is a success story. However, defective bodies sometimes survive this early screening and are allowed to live. The cultural sensibility of the time will determine what attitude a society will adopt towards these problematic lives. It can choose to mercilessly get rid of them; or, in a compassionate move, it may opt to accept and take care of them. A society which chooses to protect all human lives, even handicapped ones, bears testimony to one aspect of our humanity that raises us above the animal in us. Such a society strives to overcome natural selection by rejecting indifference and choosing compassion, care, and healing. For such a society, illness becomes an opportunity to learn from Nature's mistakes. It has been said that a genetic mutation is an experiment of Nature. This is very true: it is an occasion to discover, to know, and to understand. Better than that, it is call on what is distinctive of us in the animal kingdom: our humanity.

A hereditary mutation leading to a disease that causes childhood mortality or reduces adult fertility tends to become increasingly scarce in a population with passing generations due to the early death of children carriers and failure of adult carriers to pass them on to their descendants. Therefore if such a disease remains prevalent in a given population, it must be because the mutated gene that causes it had offered or still offer some sort of benefit for survival or reproduction in the environment in which that population lived or lives. Such a mutation can be perceived as a liability for the sick person, but it is actually an asset for the population.

One illustration is the 'S' (sickle) mutation in relation to drepanocytosis (sickle cell disease) and malaria. The mutation occurred a long time ago in the 'A' (for adult, by opposition to the embryonic or fetal) beta globin gene. This gene directs the production of one of the two (alpha and beta) proteins of hemoglobin, the internal pigment of red blood cells which captures oxygen in the lungs and distributes it as a fuel to all the cells of the body. Drepanocytosis is a potentially deadly disease; it is characterized by, among other symptoms, excruciating pain in the joints (due to the occlusion of blood capillaries by misshaped red blood cells), and severe anemia (due to extensive destruction of these cells). Malaria is an infectious disease caused mostly by *Plasmodium falciparum,* a parasite transmitted from human to human by mosquitoes' bites. The parasite multiplies in the liver first, and then in red blood cells, leading to the extensive burst of these cells, loss of hemoglobin and anemia. In its most severe form, malaria could affect the lungs and the brain leading to respiratory distress, seizures and coma. A child who inherits two 'S' copies (SS homozygote) of the beta-globin gene, one from each parent, will suffer and could die of drepanocytosis. A child who inherits an 'S' copy from one parent and an 'A' copy from the other parent (AS heterozygote) is generally resistant to severe malaria because its red blood cells are inhospitable for the parasite and impede its proliferation. In contrast a child who inherits the normal 'A' copy from both parents (AA homozygote) would be more vulnerable to the disease because his red blood cells are propitious to *P. falciparum* breeding. Malaria has been the cause of children hecatombs during evolution. Natural selection seized the chance of a random mutation in an individual's genome and, guided by blind necessity, spread the mutation, saving the population from extinction, sacrificing in the process a minority of individuals as collateral damage. The geographical distribution of malaria corresponds rather well to the demographic distribution of the S mutation. The disease is endemic in the tropics, particularly in sub-Saharan Africa where the 'S' mutation is prevalent. In these regions, AS heterozygotes being more numerous than AA and SS homozygotes, they help keep the mutation present in the population from one generation to the next.

What is true of the S mutation is also true of a variety of other changes that are apparently harmful to health in the modern ecology. If they are still present in populations, it is because, at a given period of evolution, they had helped these populations survive. A case in point is the alpha-antitrypsin gene which directs the production of an inhibitor of trypsin, a protein-cutting enzyme released by the pancreas. Some invalidating mutations in this gene cause diseases of the lungs (emphysema) and the liver (cirrhosis). Twenty percent of populations of European descent carry one mutation or another of this gene. The reason might be that in prehistoric Europe these mutations conferred protection against infectious lung diseases.

Even more enlightening is the alcohol intolerance shown by one in three East Asians (e.g., Chinese, Koreans, Vietnamese, and Japanese) who are genetically deficient in a liver-detoxifying enzyme called aldehyde dehydrogenase (ADH2). Carriers of defective ADH2 gene variants are alcohol-intolerant: after drinking an alcoholic beverage, they generally suffer from highly discomforting flushes, skin redness, increased heartbeat, and nausea. The discomfort is severe enough to discourage many of them from consuming such beverages. Could this aversion have had an evolutionary advantage in ancient periods and against what deadly disease? Probably yes: against hepatitis B virus (HBV) infection. This mother-to-child transmissible disease was endemic in ancient Asia. It ultimately led to deadly liver cirrhosis and cancer. Alcohol accelerated this fatal outcome. Those who drank it, supposedly starting at young age, died in greater number from HBV infection; those who, because of their mutated ADH2 gene abstained from it, survived. Since the mutation 'protected' against premature death, evolution conserved and spread it in East Asian populations.

Thus, in its choices of gene variations to be perpetuated, Nature is, in the long term, always 'benevolent' towards the species, notwithstanding the penalties of morbidity and early mortality paid by some individuals. *We are all the products of a positive or purifying selection of genomic mutations.* This reality is far from the 'cruel' vision of evolution propounded by some theorists.

That said, there are genetic diseases that persist in some populations because of sociological constraints that prevent progressive dilution of the noxious genes. Inbreeding is one of these constraints. In chapter 1, I alluded to the ectrodactyly (lobster toes) of the Vadoma of Southern Africa as an illustration of the consequence of inbreeding. Another example is Tay-Sachs disease, a childhood pathology characterized by progressive loss of nerve cells in the brain and the spinal cord: it occurs predominantly (1 out 4000 births) in Ashkenazi Jews who traditionally choose to reproduce within in the confines of their communities.

Over its basic structures which ensure long-term survival, the genome indulges in diversity, in genetic variations, the SNPs that I mentioned in Chapter 2. If they are not neutral, these variations offer in the midterm a wider margin of adaptation to environmental fluctuations. They collectively confer predispositions to biological vigor during a given period and in a given environment. When the context changes, these predispositions can sometimes become biological vulnerabilities. For instance, there are genetic variations that encourage the accumulation of caloric reserves in the form of fat. This caloric parsimony favored survival in the very ancient past marked by periodic famine. In modern days characterized by continuous food abundance, these variations are the principal cause of the rapid rise of obesity, diabetes and cardiovascular diseases in developed countries and wealthy classes in developing countries. In the past, rotund individuals survived, and skinny ones died. The opposite is true in the present. As for the future, if the current trend continues, it is very likely that genetic variations of caloric prodigality will be more frequent in human populations.

Similarly, genetic variations which accentuate the recapture of sodium by kidneys were a valuable asset for maintaining blood volume at an adequate level in populations with low-salt dietary traditions, among them, Central African peoples. These same variations are believed to be responsible for the surge of high blood pressure due to excessive retention of water in the bloodstream, now that these populations are eating a copiously salty diet. I believe that the popularization of salting of fish and other foodstuffs for conservation, in place of the traditional sun-drying and smoking, has compounded the problem of high-blood pressure in sub-Saharan Africa. This stands as an example of misalliance of genetics and dietary habits, a misalliance that can be corrected by public education.

Historical and geographical epidemiology allows us to trace back the conversion of biological vigor into biological vulnerabilities which environmental changes (diet, in this case) impose on organisms. Medical anthropology, as we call this new science, interrogates and investigates our past in light of the current actions of our genes. With the knowledge it provides, conditions permitting, we, as humans, can wisely change our lifestyle to 'counter' the deleterious effects of our genomic memory. Thus, behind a genetic predisposition to disease, there is often a prehistory of biological vigor. This predisposition ceases to be a fatality once we know its past history.

Genetic mutations defend us against the trials of natural selection. Variations adapt us to environmental changes. Both efficiently protect the species and its populations, the former in the long term, the latter in the midterm. Are there genomic variations that protect the individual in the short term? The answer is yes. They are called epigenetic variations. They are minor, transient and nonhereditary labels placed on genes to modify their expression in response to the immediate environmental signals. These signals are as numerous and as diverse as the environment itself. They are physical or psychological, and, I venture to say, spiritual. In so far as spirituality finds its source in, and is the product of atomic and subatomic interactions, one can speculate that it affects gene expression. Epigenetic modifications begin at conception, go on during gestation, get consolidated at birth, and adjusted throughout a person's life, improving his chances of survival and reproduction.

An example to illustrate this point: if you smoke or have smoked, do you remember how your body reacted to your first cigarette? It violently rejected the smoke, warning you a pathological danger of the activity. After a series of ignored alerts,

Chapter 4: The perversion of genetics

through epigenetic modifications, your body would silently work at rearranging the expression of your genes to detoxify your lungs and allow you to survive... until it no longer can. *Indeed, 'compassion' is a signature of the genome.*

Animal experimentation has shown that the fetus 'senses and evaluates' the quality of the food its mother eats. Through epigenetic labeling, it adjusts the development of its organs to optimize its health after birth under the same diet. An abrupt postnatal dietary change would make it inexorably ill. Where we are, what we are, what we eat, or what we think can influence ever so subtly the structure and motions of our genome. Thus, over and above the genomic modules, mutations, and variations, it is in the anticipation and experience of life, as manifested by epigenetic imprints, that the highest biological diversity unfolds.

The chemistry of genes is the same among all living beings, from virus to humans. This DNA community has allowed gene intrusions and transfers from one species to another during evolution. I have already alluded to the fact that of a good portion of our genome is microbial in origin. But microbes have not only donated their DNA to us; they share their lives with us. I previously stated that the motions of our genome, in communion with our environment, affect our wellbeing and our abilities.

Recently, there has been much talk about another environment that literally 'colonizes' us: the microbiome, the vast community of microbes that live, most often peacefully, in various parts of our body – the skin, hair, mouth, nostrils, esophagus, stomach, intestines, and vagina. These niches differ in microbial composition. They also differ among individuals. Collectively, they contribute to our physical and mental health in a variety of ways. We acquire them from the beginning of our lives, during gestation, vaginal transit at birth, lactation, feeding, and contact with our environment. Their genes are 100 times more numerous than those of our genome, making us *'walking microbial ecosystems'*. In addition to helping us digest certain foods, they also release alcohols and gases that regulate our biochemistry.

A striking example of their influence on our health is the size of our belly. Apparently, obesity or leanness of our bodies do not originate from our genes and dietary habits alone; it is also due to the panoply of germs that we carry in our stomach and intestines. Experiments have shown that an obese mouse can be rendered thin by eliminating its microbiome with an antibiotic treatment and by inoculating it with that contained in the feces of a thin mouse; and vice versa. Serious and long-lasting alterations of the microbiome have been associated with multiple diseases such as diabetes, hypertension, Alzheimer's disease, etc.[55] Thus, bacteria add more complexity to our genetic diversity.

From the simple understanding of genes, we have moved to their manipulation. This revolution began in the 1970s with molecular cloning. The technique uses enzymes and vectors (transport vehicles constructed from bacterial or viral genes). It allows one to chisel any gene, to join genetic fragments in any order, to generate artificial genes, to introduce them into selected cells and transform them into factories for the manufacture of desired proteins. The insulin used to treat diabetics and the growth hormone injected into children with delayed growth due to a congenital disability are just two examples of human proteins manufactured in bacteria by genetic engineering. This technology has been used in a variety of ways to modify plants and animals, and shortly to treat human diseases.

Drepanocytosis is undoubtedly one of the best understood genetic diseases. Because expression of the defective S-beta-globin gene occurs in the marrow of long bones (e.g., femur and tibia) which are relatively accessible, the illness was the most immediate candidate for gene therapy. U.S. medical researchers

Chapter 4: The perversion of genetics

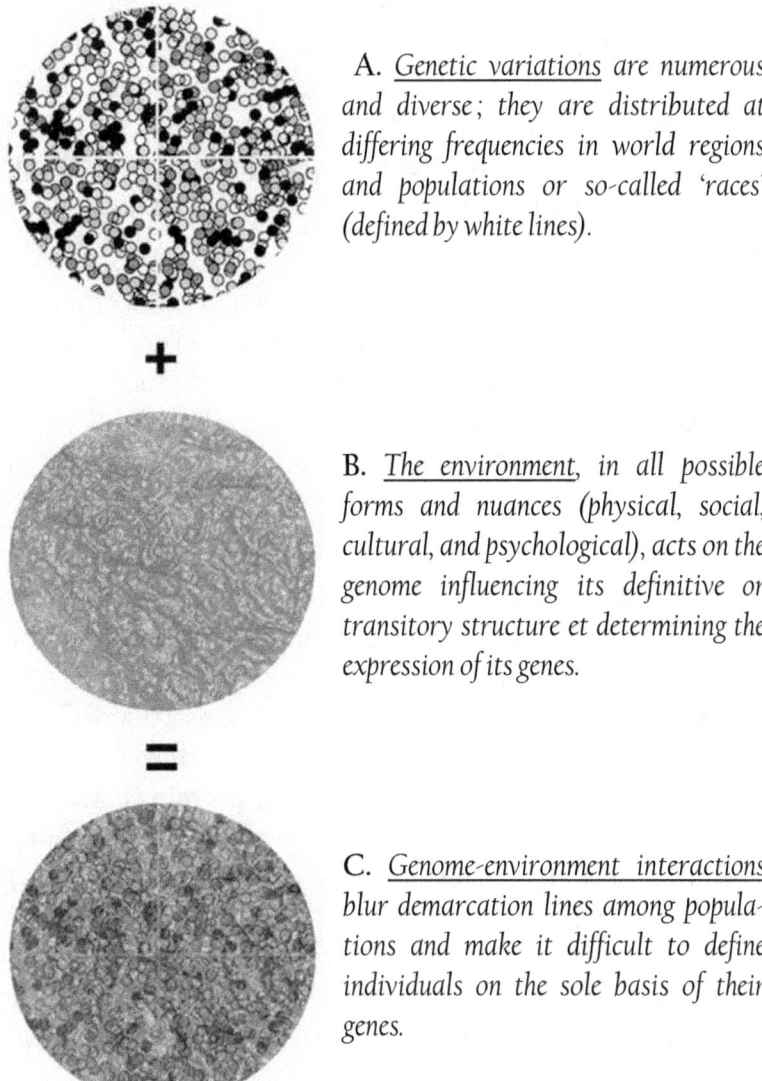

A. <u>Genetic variations</u> are numerous and diverse; they are distributed at differing frequencies in world regions and populations or so-called 'races' (defined by white lines).

B. <u>The environment</u>, in all possible forms and nuances (physical, social, cultural, and psychological), acts on the genome influencing its definitive or transitory structure et determining the expression of its genes.

C. <u>Genome-environment interactions</u> blur demarcation lines among populations and make it difficult to define individuals on the sole basis of their genes.

Figure 4.4. *The 'indivisible tandem genome-environment'*

have attempted to substitute or supplement the bone marrow of a few SS children with their marrow cells that had been modified *in vitro* to produce the good beta globin. Although promising in the short term, none of these clinical trials have been able to cure the disease for good. Efforts continue, vectors are improving. The revolutionary CRISPR/Cas9 technology (see below) is currently being tried. Early results appear promising. It is only a matter of time before hope rises again and SS children could expect to live a more or less healthy life.

A long series of metabolic, cardiovascular, cancerous or infectious, hereditary, congenital or somatic diseases are also candidates for gene therapy. Some of these, such as severe combined immune-deficiencies SCID-Xl and ADA-SCID, have been treated with some success.[56] Soon, very soon in fact, DNA will be part of our pharmacopeia.

At the writing of this essay, a millennial gene modification technology, named CRISPR/Cas9, is on the horizon. Microbiologists had noted that a bacterium immunizes itself against an invading virus (bacteriophage) by inserting fragments of this virus into its genome between repeated labels called CRISPR (*Clustered Regularly Interspaced Short Palindromic Repeats*). Coupled with a chiseling enzyme (Cas9), a guiding protein (PAM) and a recognition RNA fitting a genomic region of the bacteriophage, the system allows the bacterium to recognize the invading virus and to cut its DNA into pieces, thus destroying it. The guiding and the cutting are marvelously precise. Scientists converted this bacterial mechanism into an ingenious technology for correcting any genome after introducing its components in a target cell. The term correction in this case encompasses all its nuances, i.e., erasure, retouching, rectification, amendments, improvement. So far, the success of this technology has been astounding. With the almost daily refinements brought to it, scientists should be able to carry out increasingly bold nucleotide-specific genetic surgery in somatic and germline cells.

Chapter 4: The perversion of genetics

To what ends? To cure genetic diseases or inactivate pathogens? To 'improve' plant or animal species, including *Homo sapiens*? Lately, a Chinese scientist by the name of Jiankui He has used the CRISPR/Cas9 technology to modify the CCR5 gene in human embryos. The embryos were generated by in vitro fertilization with eggs from a lady infected by the human immunodeficiency virus (HIV); they were genetically modified and later implanted into the lady's uterus. The protein encoded by the normal CCR5 gene helps the virus enter into immune cells. Some mutations of this gene abrogate this chaperone function and people carrying them are protected from HIV infection. It is one such mutation that Dr. He introduced in the embryos. Twins born after this genetic manipulation were free of infection.[57] Success? Yes but... Sometime afterwards, a demographic survey found that, on average, people carrying the CCR5 mutation, for some unknown reason, die earlier than those without it. The finding was later dismissed. Nonetheless, it was clear that the gene did not evolve to facilitate HIV invasion (an unintended consequence), but to somehow afford some benefits to human beings.

This goes to show how ignorant we still are about the *raison d'être* of all our genes. We have learned a lot about the workings and tools of evolution, but this knowledge is incomplete. We know better the chemistry of life, but we have yet to fully understand how life acts on or responds to the Universe surrounding it. It took millennia for our genome to evolve to its present state. We must therefore exercise extreme caution when it comes to manipulating it, in germ cells (i.e., sperm or eggs) in particular lest we deleteriously affect generations to come.

Do eugenicists already dream of using the like of CRISPR-/Cas9 technology to 'perfect' the human species, to enlarge his brain and multiply his neurons and neural interconnections? Is the fiction of yesterday's novels and films closer to reality? Can we still close the Pandora's Box? For sure premature or uncalled for experiments in that line will be attempted by some uninhibited scientists. I hope that collective wisdom backed by moratoria and regulations will discourage such attempts, as it is trying to do with the potentially destructive power of the harnessed atom. We are *Homo sapiens* after all; we must strive to live up to the name.

4.5. The silence of genes

WE KNOW WHAT GENETICS says. But very often, naive or bright minds make it say what it doesn't mean. In this debate about group psychometry and the genetic heritability of its measures, I do not deny the reality or importance of genes. Genes dictate what we are, our abilities and our potentialities. However, it is when transcribed on an 'environmental parchment' that this dictation takes reality, form, meaning, and duration. The dictation may be perfect or imperfect; the parchment may meet expectations or not; it is in the fusion of the two aspects that life expresses itself. It is erroneous to think of the genome and the environment as opposing forces. Without one, the other has no meaning. *We are the dictation and the parchment merged into one in an infinite variety of nuances.* Arbitrary groupings of humans based on superficialities do not do justice to these nuances. To enrich, mix, and expand these nuances to the benefit of the species, evolution and genetics have allowed us to reproduce among ourselves, to form one unique species. Notwithstanding the eugenicists' claims to the contrary, any 'inbreeding' effort to thwart this deployment and blending of nuances is counter-genetic. Pushed to the limit, it is ultimately dysgenic. It leads to degeneration of organisms.

In the genetic lexicon, the word 'race' no longer exists; it has been expunged by the study of human genetic variations.

Chapter 4: The perversion of genetics

These variations are more numerous within groups than between groups. They are overlapping and superimposed. Genes have authorized physical differences mostly as adaptations to differences of climate and geography. These body adaptations form gradients that challenge the definition of the word 'race'. Population admixtures occasioned by the dismantling of sociological and geographical barriers will soon render this notion increasingly obsolete. Soon, we will no longer talk about our single ancestry, but about our multiple ancestries.

However, the word 'race' is not about to disappear from the vernacular vocabulary. It is an atavism imprinted in the human psyche. It stems from a need for self-assertion through differentiation, a quest for reassurance that resemblance affords. When one group wants to differentiate itself from another, it calls its own and the other group 'races'. Is there any Jewish, Arab, Bantu, or Nilotic 'race'? The term is used lightly, thrown around in all directions. Is this need for distinction embedded in our genes? Well, maybe: as a primitive reflex of self-preservation when facing a stranger, to avoid becoming a prey to a potential predator. As many things human often go, the reflex can be distorted, reinforced, and rationalized for selfish advantages.

In this regard, I recall an amusing incident which occurred to me in 1987, and which illustrates this point. I was working at IRCM at the time. Each spring, the institute organized for its employees an annual field trip, the traditional 'sugar shack', an opportunity to stuff themselves of beans and salted bacon grills, to savor maple taffy freshly squeezed from trees, to 'square dance' round and round under delightful 'country' melodies. It was by bus that we went to the 'sugar shack' that year. My wife Mujangi-Annie and our two-year-old daughter Tshanda Roxanne were with me. We entered the vehicle. Seeing the many pale faces that populated the bus, our little girl got scared. Clinging to her mother's neck, she was rejecting all the playful kindnesses of my Caucasian colleagues, until she saw Rodney Squire, a Ghanaian, sitting in the back of the bus. He extended his arms to her. She came down from her mother's knees, ran towards him, and hugged him. I stood up and, to the astonished audience, I said: "I am sorry, friends, my little girl is still a 'racist'. Give her time to get over it."

This animal heritage is still alive in all of us. However, even in animals, this reflex fades with daily familiarity. Lowering their guard, putting aside the look of mutual distrust, animals can become 'dinner guests', especially if they are of the same species. The exacerbation of distinctions within the *Homo sapiens* family is a strictly human social enterprise. Its discourse derives from a power agenda, that of entrenching into genetic fatalities physical and social differences as well as economic and military inequalities. Yet, the history of civilizations, with their rises and falls, should teach us much on the subject: the genome yes, but only in resonance with the environment. Changes in both are the only certainties that the future holds. The existential challenge for all of us as a species is to 'master' these changes.

Talking about intelligence, does it have a genetic basis? Yes, of course, it does. Like any natural aptitude, it is a product of genes. I am not in favor of the 'Blank Slate' theory, feverishly contested by Steven Pinker, the Canadian-American psycholinguist and a professor at Harvard University, in his book by the same title.[58] The theory minimizes the contribution of genes in favor of the environment. The bias derives from an aversion to, if not the fear of, the believed 'irremediable nature' of genetic traits. Although I don't espouse Pinker's views about the entrenched nature of some gender or "racial" differences, I do agree with him that they originate from genetic variations.

We know that spontaneous genetic mutations that occasionally occur in humans can induce abnormal changes in behavior or the perception of reality. Think of the microdeletions

Chapter 4: The perversion of genetics

of chromosome 15, which, when inherited from the father, induces the Prader-Willy syndrome, a disease characterized by massive obesity due to an insatiable appetite. These same microdeletions, inherited from the mother, induces the Angelman's syndrome (nicknamed 'joyful doll syndrome') which is marked by physical thinness and a joyful mind disposition. As there are familial predispositions – which are possibly genetic in origin and hereditary in nature – to diabetes, obesity, high blood pressure, and cancer, so too are there predispositions to depression, schizophrenia, autism, and homosexuality, etc.

What is true of the physical and the mental is also true of this unique ability for humankind to conceive, abstract, anticipate, and invent: of intelligence. Our objection therefore does not concern the genetic nature of this ability or its heritability. It concerns two major points: the definition of intelligence and its comparative measurement between human groups.

Naming something does not make it an entity. What is called intelligence is a concept; and, as such, it escapes the restrictive limits of a definition. We can see its manifestations, effects, actions, and reactions, but we cannot capture it as a whole in a text or a test. A distinctive feature of the man is that ideology (the world of ideas) is an integral part of his environment. Only through the magnifying glass of the dominant ideology of his society can a person evaluate the intelligence of another person. Depending on their evolution in particular environments, societies may differ in their existential priorities. Some societies may value more spiritual growth (e.g., Tibetans), others social harmony (e.g., Africans), or co-management of Nature (e.g., Amerindians), still others the domestication of Nature and conquering materialism (e.g., Westerners). All these societies will express their intelligence in a particular way. With the contraction of our planet thanks to modern means of communication and mobility, these various forms of intelligence compete among them for relevance. The 'intelligence' that promotes material conquest currently has the upper hand from an economic and military points of view. Fascinated by it, all modern societies strive to express it to the best of their environment. As Angela Saini puts it in her recent book entitled *"Superior, the return of race science"*:[59]

> "There's an implicit assumption that higher productivity and more mastery over nature, the presence of settlements and cities, are the marks of human progress, even of the evolution of humankind. The more superior we are to nature, the more superior we are as humans."

There are no specific genes of human intelligence as there is none of goodness or beauty, compassion or cruelty. Intelligence results from interactions among the products of a multitude of genes expressed in accordance with the needs for survival, procreation, and creation in a given environment.

It is strictly against this background that one must discuss the validity of IQ tests. The latter do measure something, but this thing is not intelligence (which is not a thing), but specific abilities according to parameters established by the tester. *The test cannot but be a mirror, and the result, a reflection of the "man with the ruler".* Regardless of all the tester's efforts of impartiality and indifference, the test will always consist of confronting a testee with the tester's yardstick. The testee's response will depend subconsciously on his or her acceptance of this confrontation. It is difficult to rid the IQ tests of this emotional element, especially in a society like the United States, with a long history of aggression and resistance between groups of different epidermal coloration.

IQ tests have more than 100 years of history since the French psychologist Alfred Binet (1857–1911) developed them to help children showing schooling difficulties. Since then, the controversy surrounding them has not lost any of its ferocity, especially since eugenicists and 'race scientists' took hold of them and turned their results into a hereditary fatality. A care-

Chapter 4: The perversion of genetics

ful review of the literature on this subject has revealed numerous misuses of scientific terminology aimed at explaining confusing data and supporting dubious theses. For the most scathing denial of the validity of IQ tests, one should read the book entitled *The Mismeasure of Man*,[60] written by Stephen Jay Gould, the same scientist who formulated the theory of punctuated equilibriums during /evolution to which I alluded in Chapter 3. This scholarly historian used the data and analyses of the 'IQists' themselves, and systematically pinpointed the inconsistencies they contained, irrefutably debunking their conclusions.

Since genetic differences between human groups are insignificant compared to the genetic variability among individuals within groups, and since genes dictate intellectual faculties, it follows that these faculties are variable among individuals, but not among groups. Any difference among groups in the measurement of one or other of these faculties is invariably an artifact of the environment (in the broadest sense word: climate, nutrition, health, family, education, freedom, opportunity, culture). Changing the context for the best and the highest possible number of people will not level the faculties, but it will allow their full expression to their variable amplitude and diversity. Sadly, this offer and promise made to us by our genes are too often frustrated by pseudoscientific theories that attempt to justify and perpetuate entrenched social hierarchies.

CHAPTER 5

LOOKING TO THE FUTURE

5.1. Genetics must not be a religion

AT TIMES, THE TONE of this book may seem somewhat polemical (from the Greek *polemos* = war). I fully assume this remark without apology. However, if there is a war that I wanted to wage through these lines, it is in no way against the desire to know and explain, but resolutely against the impudent and abusive utilization of science – genetics, in this case – to defend and spread preconceived ideologies. Today, science has authority over our vision of reality. Scientists have a preponderant voice on many subjects. In the enthusiasm over their achievements, some of them ignore the temperance and humility that ought to be theirs in the face of the immense ignorance yet to be conquered.

For its part, the lay public is often complicit with those scientists inclined to engage in premature speculations and excessive extrapolations. The proclaimed certainties that yesterday peoples expected from their priests, pastors, imams, mediums or other marabouts, they now expect them from professors in white blouses. They eagerly want genes to reveal to them the 'hidden truths' inscribed in their DNA. They implore scientists, the celebrated "readers of genetic epistles" to decipher for them the destiny that the sacred genetic annals hold for them. They want to know whether they will live in heaven or hell on Earth.

With the advent of social media, loquacious bloggers have been able to recruit gullible multitudes to which they feed on a daily basis their distorted pseudo-scientific views on genetic differences among human groups. In his recent book entitled "*Skin Deep*"[61] about the renaissance of 'race science', the London-born South African author Gavin Evans forcefully takes issue with the pronouncements of the most illustrious of these media preachers. I refer interested readers to the book.

As a scientist, I'd rather put blame on those of my colleagues who, by intent or negligence, nourish these ideological fanatics. Intellectual honesty demands that geneticists resist the temptation to play high priests or prophets; that they realistically circumscribe their discoveries, publicly accept the unknowns still surrounding them, and vigorously contradict any distortion that public opinion – especially the popular press – so often imprints on them. Science has its vocabulary, but not everyone who uses this vocabulary understand its meaning and definition; People brandish the word 'DNA' to explain their habits or attitudes. Depending on circumstances, they impart a nuance of sublime nobility or fatal wretchedness to the words 'hereditary' and 'genetic'. With utter contempt, they attach the latter epithet to their invective. To my surprise and amusement, I once heard the wife of a friend of mine in a bout of anger call her husband 'genetically poor'. In the woman's mind, her man's economic destiny was irremediably sealed in his genes.

What could be more tragically fatal than a defect in the 'thinking machine?' The taxonomists of intellectual faculties among human groups have made many concessions to the individual. They have widened bell curves and refrain from presuming of the intelligence – and lack thereof – of anyone, even if that person belongs to an ancestral group they consider to be less intelligent. "We measure 'racial' average abilities, not individual ones!" they clamor defensively. The claim is a mere subterfuge, no less aggressive and hurtful than a frontal attack.

Chapter 5: Looking to the future

Perniciously, it submerges the subconscious mind of people and subverts their views of ancestral groups with unconscious negative or positive prejudices, leaving them with a warped feeling of fairness and innocence.

> "You, you're not like the others!"
> "It's really amazing how smart that girl is!"
> "I'm friendly to everybody, even to people of color."

Such phrases, uttered by a non-African to an African, hide a denigrating and contemptuous view of all Africans. The perversity must be exposed to broad daylight and recalled to our collective consciousness. As I have said, genetically speaking, individuals are very different and populations very similar. Steven Pinker puts it well in his previously mentioned book:

> "The differences in skin and hair that are so obvious when we look at people of other races are really a trick played on our intuition. Racial differences are largely adaptations to climate. (..) The parts of the body that face the elements are also the parts that face the eyes of other people, which fools them into thinking that racial differences run deeper than they really do."[62]

A joke of very bad taste indeed for those on the wrong side of the equation! The search for 'intelligence genes' is, in my opinion, a futile exercise. Selective sweeps in and around intelligence genes that would definitely distinguish human populations will forever await their honest discoverers. Rather than indulging in hollow speculation, when it must, the inquisitive mind, should direct his or her gaze at the individual, at his or her physical and mental abilities, not at the 'racial' group within which society would like to cast him or her. Dr. Martin Luther King Jr. (1929-1968), forcefully expressed this ideal in his historic message concluding the Washington March for Jobs and Freedom on August 28, 1968:

> "I dream that my four grandchildren will one day live in a nation where they will not be judged by the color of their skin, but by the content of their character."

This open-minded attitude is possible; it happens often enough; but only when the anxieties of personal or collective survival subside. Indeed, the ranking of peoples and populations is fundamentally a biological reflex, a remnant of the evolutionary times when survival and procreation largely depended on the status one held within the band or the tribe, on the dominance of one tribe or band over others. This reflex still sneakily guide human behavior. As creatures of reasons and pretenses, humans try to rationalize it by wrapping it in self-serving theories, yesterday theological, today genetic.

The hidden intention of these theories remains privileged access to food and sex. It is not so long ago that men killed each other to secure a nourishing territory, to capture women as spoils of war and inseminate them to beget an abundant progeny. In today's jargon, peoples speak of 'preserving their way of life'; they frown with apprehension over the mixing of gametes between members of groups with a differing vital interest, short of opposing it violently or legally. Through history, sex between members of rival groups, be they called classes, castes or 'races', has been condemned or criminalized in many societies (e.g., India, South Africa, the United States). Trans-group rape has always been an act of domination or vengeance. It physically and psychologically ravages female victims, and destroys the psyche of their men, burdened by the guilt of not having been able to prevent it.

To safeguard their statutory privileges and justify social inequalities and enslavements, members of dominant groups always vilify those of dominated groups. They always try to clear their conscience of any sense of guilt by invoking one or another of the theories in vogue, sacred or secular. As the German philosopher Arthur Schopenhauer (1788–1860) summed it up in one of his powerful aphorisms:

> "Every miserable fool who has nothing at all of which he can be proud, adopts, as a last resource, pride in the nation to which he

Chapter 5: Looking to the future

belongs; he is ready and happy to defend all its faults and follies tooth and nail, thus reimbursing himself for his own inferiority."

Thus, all denials notwithstanding, comparative measurements of intellectual faculties of human groups are intended to evaluate the human relevance of these groups. We who know better must expose its instinctive motivation hidden behind proclamations of divine will or scientific curiosity. We must do so clearly, loudly, and in increasing numbers, if we are to avoid catastrophic amplifications of this instinct which led to the apartheid, slavery, and recurrent genocide that pave our history.

Through his mastery of the electron, the atom, and DNA, man domesticates Nature and life a little more every day. This newly gained power gives him a growing capacity for mass destruction of his environment and all life in it. Faced with this reality, the survival of the fittest, as the supreme law of natural selection, must be reinterpreted. Soon man will cease to be a mere product of evolution; he will become its agent. The mastery could lead to our collective survival or suicide. Because he is *Homo sapiens* endowed with unique survival genes expressed in his brain, man can consciously contradict some the instincts that have guided him during past evolution. He can willfully choose tolerance of differences and temperance in violence as new mechanisms of permanence. He can contain his most extreme reflexes of distinction and distancing and replace them with impulses of acceptance of the other, as a companion, accomplice, or competitor perhaps, but not as a mortal enemy. For sure, he will occasionally revert to the beast in him, to eruptions of murderous madness, to mass ethnic cleansing of the kinds witnessed in Namibia, Germany, Turkey, Rwanda, Kosovo, and Congo during the last two centuries. That these sad events can still happen and risk happening at any time testifies to the fact that man is still a child that must evolve and mature. He must learn not to fear a different other, to be sensitive to his need and pain. He must surpass and transcend this fear if the human species is to survive. Genes expressed in his brain will remind or command him to do so. The extinction of the human species, if it occurs, will probably not come from man, but from a cosmic cataclysm or a biological disaster beyond his control.

Empty dream or wishful thinking, some may say. Well, maybe. However, if over time the evolution of genes has led to consciousness, why can't it lead to wisdom in time? The future survival of humankind will result from the accumulation of genetic variations that predispose to such wisdom, that contradict our current atavisms. *Indeed, the first mission of our genes as Homo sapiens is not to engrave our fatalities but to incarnate our possibilities.*

Science prides itself of its objective and impartial look on Nature. As biology reveals its marvels of organized complexities, science attributes them all to 'Chance and Necessity', to use the telling title words of the book by Jacques Monod (1910-1976), the French biochemist, winner of the 1965 Nobel Prize of Medicine.[63] Evolution is implicitly taken as purposeless force driving the march of the Universe and all in it. Thus, in view of past vicissitudes of life on Earth, our claim to permanence may seem somewhat presumptuous. As a life form, we may disappear at any moment. Our hope of permanence is ultimately unscientific. Evolution, it is said, has no fixed direction. The French Jesuit and paleontologist Pierre Theilard de Chardin (1881-1955) had been viciously maligned for suggesting the contrary in his philosophical assay entitled "*Le phénomène humain*";[64] for proposing that Life is a journey towards increasing 'conscientization' which will culminate at the Omega point (pre-symbolized by Jesus, the Christ, in his view). But was he completely wrong? Is evolution as blind as we biologists will have it? To some, the question may seem subversive, especially from a scientist like me. But to me, it is worth asking.

In his last popular science book, Christian de Duve, (1917-2013), the Belgian biochemist, winner of the 1974 Nobel Prize of Medicine, borrowing from the Judeo-Christian mythology of creation, called natural selection the *"original sin"* of evolution; it had permitted the deployment of survival instincts for the pursuit of personal or group benefits, often at the expenses of all others. These instincts, he writes, are the natural foundations of discrimination, injustices, crimes and wars. However, not giving to despair, he adds:

> We are indeed of all living beings on Earth the only ones that are not slavishly subject to natural selection. Thank to our superior brains, we have the ability to look into the future and to reason, decide, and act in the light of our predictions and expectations, even against our immediate interests, if need be, for the benefit of a later good. We enjoy the unique faculty of being able to act *against natural selection.*[65]

Since this 'original sin' is forever embedded in us by natural selection, we cannot count on our genes to change that. We are responsible for our own biological future. We will have to determine and direct our own evolution. Our redemption or damnation will come from us, and us only.

5.2. Africans beyond genetics

THE TRAGIC SPECTACLE of the African world and its diaspora for the last four centuries has made it the ideal target of craniometrics and psychometrics. Many felt the need to explain this state of affairs and, through these explanations, to naturalize it and cement it into biological inevitabilities. We honestly cannot ignore or underestimate the technological gulf that currently separates Africa and the West or the East. At this era when the last two worlds explore interstellar spaces, Africa *"mmusomba wuota munya"* (is lazily sunbathing), as my grandfather Nkunda Lubumbashi used to say in his moments of spite. Will Africa ever catch up with the rest of the world? Will Africans forever remain on the sideline of humanity?

To the world, the word 'African' has become synonymous (wrongly, I maintain) with misery, disease, illiteracy, even stupidity. So much so that some peoples of the continent – e.g., Egyptians, 'Maghrebins', Sudanese, or Ethiopians – distance themselves from the word, silently denying that it applies to them. The current invasion of Europe by waves of desperate migrants from sub-Saharan Africa and their repeated drownings under the waters of the Mediterranean Sea can only question consciences and bring back to mind the famous poem by the British writer Rudyard Kipling (1865–1936):

> Take up the White Man's burden
> Send forth the best ye breed,
> Go bind your sons to exile
> To serve your captives' need;
> To wait in heavy harness,
> On fluttered folk and wild
> Your new-caught, sullen peoples,
> Half-devil and half-child.
> ...
>
> *The White Man's Burden* (1903).

Today, Africa is looked at with pity if not indifference or disgust.[66] His deaths count in the millions; they do not count; they no longer move. My native Congo has lost more than 6,000,000 souls in the last 20 years. At the writing of these lines, the savanna landscape of my ancestral Kasai province is littered every day with mass graves in which lie, jumbled disrespectfully, the unnamed victims of the carnage that is taking place at this moment. Congolese speak of it as of a trivial event; they seem undisturbed by it. Soon, they will lose even the memory of it. What about the rest of the world? Pathetic! *Mawa ee!* as goes the Lingala exclamation. Honestly, will African lives matter one day?

Chapter 5: Looking to the future

In this conclusion, I will dispense with the real or invoked causes of the current state of the African world. By rehashing them, I may turn them into excuses. Many have tried. A recent example is Jared Diamond, the geographer biologist and essayist at the University of California at Los Angeles. He lists, as causes, geography (including climate and its associated evils), late onset of plant and animal domestication and writing, a shorter history of institutions capable of managing complex societies. He proposes that these factors, singly or collectively, could have contributed to Africa's underdevelopment.[67]

On this point, I beg to differ. Like many authors before him, Diamond subtracts Egypt from the continuing history of Africa. All the factors he enumerates (institutions, agriculture, writing, etc.) were met in ancient Egypt, contributing to its material and cultural wealth, millennia before the rise of Rome and Athens. If there is a deficit of these factors in today's sub-Saharan Africa, it is not because Africans lack a history, but because, during their journeys and struggles for survival in various habitats of the continent, some of them inhospitable, they have lost the memory of this history.

As I metaphorically stated in an earlier essay on the sociobiological analysis of the state of affairs in the DRC, *'History and Culture are the gametes of a Nation'*.[68] Without them, strong institutions cannot sprout, rise again, and endure. In my opinion, all the developmental factors cited by Diamond are mere circumstances and environment. They can be modified to some extent, and quite quickly, using modern tools, home-made or borrowed. What Africa needs is enlightened and determined leadership, men and women of vision and courage, fruits of a culture that values, promotes and honors these talents.

Thus, it is futile for Africans to engage in historical finger-pointing and guilt trip. It is far more productive for them to search for ways to do more than survive, to live fully and fruitfully. In this pursuit, they have to exercise *'pioneering clairvoy-*

ance': a lucid examination of the competing realities of the modern world, coupled with a will to dominate and convert them into advantages. Since we – you and I – are persuaded that the current condition of Africa is not a genetic fatality, we should explore paths towards its rehabilitation and consider foretelling signs of its renaissance.

The barriers that once isolated Africa have been dismantled; the handcuffs that once chained its limbs have been broken. But the continent has yet to unburden itself of the lingering memory of this isolation and these chains, to free itself from the insidious complexes of prisoner and servant. Many Africans have achieved this individually. They have eroded the myth in some small ways; but they shouldn't be content of their achievement as long as Africa as a whole still struggles under the yoke of these complexes, as exemplified by the unending reliance of so many of its States on foreign assistance for their security (e.g., Central African Republic, DRC, Mali) and governance (e.g., public service, elections). One can't help wondering whether the current African culture is well tuned for survival in the modern environment of multifaceted demands and competitions. The question isn't meant to implicate the genetic pool of Africans; it certainly does interrogate the way this pool is nourished and maintained. Yes, genes must be well fed if they are to express their full potential. It is well known that nutrition influences the size and health of a progeny; there is no doubt that it also impacts its intellectual performance. To say that diseases weaken the body is a tautology; that they can, in some cases, leave long-lasting noxious sequels in the brain, is more than mere conjecture.

Feeding the body is only a first step in the liberation of the genetic potentialities of a people. Nourishing the mind is even more important. A good food for the mind consists of quality education for all, especially the youth. It means the acquisition and mastery of practical tools (technologies) for survival in this modern world of unbridled changes; it means the development of mental capacities to 'dream' of a better world (ideas and ideals) and work at its materialization through adaptations and innovations. Once it has achieved these goals, Africa could then dialogue with other continents on an equal footing or, failing that, successfully sustain their competitions. Critical education is the only weapon against outdated beliefs, sickly religiosity, debilitating fatalism that still poison the mental universe of Africans. It is the only weapon that can free them from the brutal dictatorships and shameless kleptocracies that still imprison them here and there.

The culture of a people is an adaptive strategy for collective survival in a given environment. It is the product of a long history of interactions between the people and their environment. When the living context changes, some aspects of culture could be maladaptive. They may oppose or delay beneficial changes needed to ensure continuous survival of the collective. One aspect African culture that I have repeatedly questioned is the concept of primogeniture, of 'power to the oldest' that permeates African hierarchies.[69] This concept must be revisited. It stifles the emergence of young 'parallel thinkers' and geniuses. It muzzles them, when it doesn't ban them, broadly branding them as 'witches'. Leadership must not be a birth right (aristocracy), nor an old-age right (gerontocracy); it is the right of the most competent (meritocracy), no matter his age and lineage.

In my above-mentioned essay on the sociobiology of the DRC,[70] I made an impassioned plea for 'more power to the younger generation:

> The youth must respect and honor his elder, but he must never succumb to a paralyzing obsequiousness to him. He must question and challenge him. Indeed, the authority of the elder is relative and his knowledge contextual. An elder may successfully cure an illness using a traditional remedy from the local flora; but, he could be helpless in front of the keyboard of a microcom-

puter. The elder's knowledge is a heritage of a life spent in an environment that is obsolete in many respects. In the here and now, it can be useful or useless; it can teach something or nothing; it should be remembered, but not necessarily followed. For it is the product of fixed neuronal modalities leading to stereotypical actions and reactions. Broader neuronal flexibility is an endowment of youth. The youth, if properly educated, is by far better equipped to generate survival strategies fit for the challenges of his time and environment. (...). He must be encouraged to create new knowledge, to surpass what has been bequeathed him by his elders, to always ask this simple question: "Can I do better?" It is the question of the innovative mind, a mind that obsessive respect for the frozen knowledge of the past could asphyxiate. The role of the elder is not to guide the youth into the future, but to carry him on his shoulders that he may better scrutinize this future before entering into it. For Time, in its inexorable course, continually erases old horizons to open new ones that only the youth can see. The role of the youth is to firmly rest on these carrying shoulders, while he extends its leafy branches towards the sky of his future, aware that these shoulders are the roots which, buried under a past soil, feed him the sap that keeps him alive.

From a rich pool of well-attended genes variably expressed and recombined, will emerge from Africa inspired and inventive, audacious and valiant minds – messiahs, reformists, or leaders – who will rid the continent of past burdens and obscurantisms, leading it to new and radiant vistas, where atavistic instincts of individual survival will cede the reins to intuitions of the spirit, circumventing the blind traps of natural selection in a perpetual quest for a better tomorrow for a majority of its people.

The road will be long and hard, for sure. The big picture of this noble march will be stained with spotty events of egoistic pursuits, gratuitous cruelties, and painful setbacks. But the march will go on, driven by faith, hope, and unfettered intelligence. From my perspective at least, there is evidence that this

Chapter 5: Looking to the future

change of mentalities is already under way, especially among young Africans. Africa, some say, is the continent of tomorrow. The next Einstein may be an African.

Notwithstanding bell-shaped IQ curves.

NOTES AND REFERENCES

[1] Contrary to modern thinking, according to traditional African classification of species, man does not belong to the animal kingdom. However in the totemic spirit of tradition, man can share similar temperamental traits with animals. In Tshiluba the animal kingdom is called *'bukua nyama'*; the human kingdom, *'bukua bantu'*.

[2] Exceptions are trans-specific reproduction experiments forced by humans; their progeny, such as the mule (stallion × ass), the tigra (tiger × lioness), the ligra (lion × tigress), are generally infertile or sterile. This type of crossing would have occurred naturally in the past between *Homo sapiens* and *Homo neanderthaliensis*.

[3] Hiernaux, Jean. *Luba du Katanga et Luba du Kasaï (Congo): Comparaison de deux populations de même origine*. Bulletins et Mémoires de la Société d' Anthropologie de Paris, tome 6, XIe série, pp. 611 à 622, 1964. Error! Hyperlink reference not valid.

[4] Torday, Emil. *The principles of Bantu marriage*. Africa: Journal of the International African Institute, 2:255-290, 1929.

[5] Lembe-Masiala, Nathalis. *Le káandu chez les Basolongo du Bas-Congo (RDC)*. PhD Thesis, Langues et Cultures africaines, Ghent University, 2007.

[6] In Tshiluba, the term used to designate an albino is *'Tshitokatoka'*. It has a pejorative connotation and, in its most cruel sense, could be translated into 'White Monster'. At the instigation of the Mbuji-Mayi Catholic diocese, an elegant and much more benevolent appellation has recently been proposed: *'Mutoka wa kuetu'* (The White of our land!)

[7] Godbey, AH. *Ceremonial spitting*. The Monist 24:67-91, 1914.

[8] We now know that the sharing of biological fluids amounts to the sharing of microbes that these fluids contain. As we shall see, in Chapter 4, these microbes influence our physical and mental development to some extent.

[9] Edelstein, Stuart J. *The sickled cell: from myths to molecules*. Harvard University Press, 1986.

[10] Ibid.

[11] Goblet-Vanormelingen, Véronique. *La maison du MBOMBO: Rite thérapeutique pour les enfants à hauts risques dans le Zaïre rural*. Social Science & Médicine 37:241-252, 1993.

[12] Lawal, Musediq Olufemi. *Cultural conception and management of Sickle cell anaemia among the Yorùbá in Osun State, Nigeria*. PhD thesis, Department of Sociology, University of Ibadan, 2012

[13] De Smet, M, and De Visscher, M. *Contribution à l'étude de l'endémie goitreuse des Uélés (République du Congo)*. 1960.

[14] Kabasele Lumumba, François, Nkongolo, Sylvain, and Anganga Miki, Marcel. *Naissances insolites en terre africaine*. Karthala Editions, 2013.

[15] Weeks, John H. *Among the primitive Bakongo: a record of thirty years' close intercourse with the Bakongo and other tribes of equatorial Africa, with a description of their habits, customs, & religious beliefs*. JB Lippincott, 1914.

[16] Bouquet, Armand. *Féticheurs et médecines traditionnelles du Congo (Brazzaville)*. 1969.

[17] Mendel, Gregor. *Experiments in plant hybridization (1865)*. Verhandlungen des naturforschenden Vereins Brünn, 1996.

[18] Darwin, Charles. *On the origin of species by means of natural selection, or the preservation of favoured races in the struggle for life*. John Murray, 1859.

[19] Judson, Horace Freeland. *The eighth day of creation*, Jonathan Cape,

Notes and references

1976.

[20] Watson, James D, and Crick, Francis HC. *Molecular structure of nucleic acids.* Nature 171:737-738, 1953.

[21] Eldredge, Niles and Gould, Stephen Jay. *Punctuated equilibria: An alternative to phyletic gradualism.* In Schopf, Thomas J.M. (ed.), *Models in paleobiology.* Freeman, Cooper & Company, pp. 82-115, 1972.

[22] Barker, David J. *The fetal and infant origins of adult disease.* British Medical Journal 301:1111, 1990.

[23] However, an alternative theory, supported by many scientists, is that these molecules came from Space, brought about by meteorites bombarding the Earth's surface at its origins. Star dust: would we draw our origin from the dust of stars?

[24] Gilbert, Walter. *Origin of life: The RNA world.* Nature 319: 618, 1986.

[25] Paleontology is a constantly moving field. As more humanlike fossils are being excavated and dating methods becoming more sophisticated, the approximation of the appearance time of Homo sapiens is bound to change. For instance, the finding in 2017 of human remains in the Jebel Irhood cave of Morocco, pushes back the birth of Humanity to 300 years ago. Even as the estimated time range expands, and debate over the primordial locations lingers on, the broad consensus among scientists is that this these locations are definitely African as corroborate by genetic studies.

[26] Sturm, Richard A, and Duffy, David L. *Human pigmentation genes under environmental selection.* Genome biology 13:248, 2012.

[27] Lewontin, Richard C. *The apportionment of human diversity.* In *Evolutionary Biology*, edited by Dobzhansky, T., Hecht, M. K., and Steere, W. C. Springer, New York, 1972.

[28] Ibid.

[29] Edwards, Anthony William Fairbank. *Human genetic diversity: Lewontin's fallacy.* BioEssays 25:798-801, 2003.

[30] The sequencing of the genome of James Watson who, with Francis Crick, co-conceived the double helix model of DNA, has revealed that he is 16% African. Ironic for a scientist who has questioned the intelligence of Africans (see chapter IV).

[31] Hefny, Mostafa. *I am not a white man but the U.S. Government is forcing me to be one*. Red Sea Press, 2019.

[32] Galton, Francis. Hereditary genius: An inquiry into its laws and consequences. Macmillan, 1869.

[33] "What I have written is written." John 19 :22-22

[34] Gobineau, Arthur. *Essai sur l'inégalité des races humaines*. Didot frères, Hanovre, Rumpler, 1855.

[35] Shockley, William B. *Shockley on eugenics and race: the application of science to the solution of human problems*. Scott Townsend Publishers, 1992.

[36] Shurkin, Joel N. *Broken genius: The rise and fall of William Shockley, creator of the electronic age*. Palgrave Macmillan, 2006.

[37] The public mistreatment of Sarah Baartman and Ota Benga was comforted by a generalized underlying sentiment on the part of Caucasian spectators that these Africans were not 'really human', at least not as 'human as we are'. Sarah and Ota are victim-symbols of this enduring superiority complex. Africans today and forever need to keep these symbols alive as reminders of the dangers of eugenics and related ideologies. The same way President Nelson Mandela requested and obtained in 2002 the return of Sarah's remains from France to South Africa, so must DRC leaders demand the return of Ota's from the U.S. to erect a monument upon his new burial ground on the soil of his ancestors.

[38] Rushton, J Philippe. *Race, evolution, and behavior: a life history perspective*. Transaction Publishers, 1996.

[39] Herrnstein, R.J., and Murray, C.A. *The Bell Curve: intelligence and*

class structure in American life. Free Press, 1994.

[40] Rushton, op. cit., chapter I.

[41] https://www.theguardian.com/world/2019/jan/13/james-watson-scientist-honors-stripped-reprehensible-race-comments

[42] https://www.nytimes.com/2019/01/01/science/watson-dna-genetics-race.html

[43] Ibid.

[44] Mbikay, Majambu. *Signposts of hopes and illusions.* Lulu.com, 2016.

[45] Diop, Cheikh Anta. *Civilisation ou barbarie: anthropologie sans complaisance.* Présence africaine, 1981.

[46] Fourche, JA Tiarko, and Morlighem, Henri. *Une bible noire: cosmogonie bantu.* Les Deux Océans, 2002.

[47] Tempels, Placide. *La philosophie bantoue.* Lovania, Élisabethville, 1945.

[48] Rushton, op. cit., chapter 4.

[49] Ibid.

[50] Gregory, M. D., Kippenhan, J. S., Eisenberg, D. P., Kohn, P. D., Dickinson, D., Mattay, V. S., Chen, Q., Weinberger, D. R., Saad, Z. S., and Berman, K. F. *Neanderthal-derived genetic variation shapes modern human cranium and brain.* Scientific Reports 7:6308, 2017.

[51] Rushton, op. cit., chapter 5.

[52] Tramo, Mark Jude, Loftus, WC, Stukel, TA, Green, RL, Weaver, JB, and Gazzaniga, MS. *Brain size, head size, and intelligence quotient in monozygotic twins.* Neurology 50:1246-1252, 1998.

[53] Perkins, Jessica M, Subramanian, SV, Davey Smith, George, and Özaltin, Emre. *Adult height, nutrition, and population health.* Nutrition reviews 74:149-165, 2016.

[54] Colle, Pierre, and Van Overbergh, Cyrille. *Les Baluba (Congo Belge).*

A. Dewit, 1913.

[55] Morand, Jean-Jacques. *Le microbiote intestinal: un organe à part entière.* Médecine et santé tropicales 27:10-10, 2017.

[56] Ginn, Samantha L, Alexander, Ian E, Edelstein, Michael L, Abedi, Mohammad R, and Wixon, Jo. *Gene therapy clinical trials worldwide to 2012–an update.* Journal of Gene Medicine 15:65-77, 2013.

[57] Cyranoski, David, *Genome-edited baby claim provokes an international outcry.* Nature 563:607-8. Dr. He was later arrested, tried and sentenced by a Chinese court to a 3-year imprisonment for having conducted this experiment without ethical approval.

[58] Pinker, Steven. *The blank slate: the modern denial of human nature.* Viking, 2016.

[59] Saini, Angela. *Superior: the return of race science.* Beacon Press, 2019

[60] Gould, Stephen Jay. *The mismeasure of man.* WW Norton & Company, 1996.

[61] Evans, Gavin, Skin Deep: Journeys in the divisive science of race. Oneworld Publications, 2019)

[62] Pinker, op. cit., chapter 8.

[63] Monod, Jacques. Le hasard et la nécessité. Essai sur la philosophie naturelle de la biologie moderne. Éditions du Seuil, 1973.

[64] Teilhard de Chardin, Pierre. *Le phénomène humain.* Éditions du Seuil, 1956.

[65] de Duve, Christian and Patterson, Neil. *Genetics of original sin: the impact if natural selection on the future of humanity.* Éditions Odile Jacob, 2010.

[66] President Donald Trump couldn't help himself when, on January 12, 2018, he treated African States of 'shithole countries'. Uttered be-

hind closed doors, these undiplomatic qualifications nonetheless translate a widespread view of Africa, of sub-Saharan Africa in particular.

[67] Diamond, Jared M. *Guns, germs and steel: a short history of everybody for the last 13,000 years*. Random House, 1998.

[68] Mbikay, Majambu. *Demain le Congo*. Lulu.com, 2012.

[69] Tempels, op. cit.

[70] Mbikay, 2012, op. cit.

LEXICON
OF RECURRENT GENETIC TERMS

Allele – Each one of all possible versions of a gene.

Chromosome – Cellular particle mostly made of *DNA* that is passed on from cell to cell, from parent to child.

Diploid – Is said of a cell that contains two copies of each chromosome, e.g., human *somatic* cell.

DNA – DeoxyriboNucleic Acid, the chemical blueprint that constitutes the material of heredity in microbes, plants, and animals including man. It is usually made of two strands of complementary *nucleotides*. Some viruses have *RNA* for normal blueprint.

Double helix – A pair of complementary *DNA* chains, linearly winding around each other.

Drepanocytosis – also known as Sickle Cell Disease (SCD). It is a genetic disease characterized by severe blood deficiency in oxygen-carrying hemoglobin due to the destruction of malformed (sickled) red blood cells.

Enzyme – Protein that accelerates a chemical reaction in a living beings

Epigenetic – Is said of a nonhereditary modifications of *DNA* that can affect its activation and function. The modifications are erased in *gametes* and restored in *zygotes*.

Evolution – Mechanism whereby species change and are selected by Nature for fitness.

Exon – Segment of a gene whose corresponding transcript remains within the functional *RNA*, after *splicing*.

Gamete – Cell produced from a germ cell in the testis or the ovary and characterized by a *haploid genome*. Spermatozoa (sperm cells) and ova (eggs) are gametes.

Gene – Segment of *DNA* containing information for a specific function.

Genetic pool – Sum of all the genes and their variations in given population.

Genome – Sum of all the genes contained in a cell.

Genotype – A genetic type characterized by the presence of specific *variation*(s)

Haploid – Is said of a cell that contains a single copy of each chromosome, e.g., a *gamete*.

Heterozygote – A person who carries two different *alleles* of a *gene* on a pair of *homologous chromosomes*.

Homologous – Is said of *chromosomes* or *genes* that carry similar (not necessarily identical) hereditary material.

Homozygote - A person who carries two identical *alleles* of a *gene* on a pair of *homologous chromosomes*.

Intron – Segment of a gene whose corresponding transcript is spliced out of the primary *RNA*, in the process of formation of a functional *RNA*.

Meiosis – Cellular division leading to the formation of *haploid* gametes.

Mitochondria – Intracellular organelle dedicated to the production of the chemical energy that drive biochemical reactions. Mitochondria presumably originated from colonizing bacteria engulfed by ancient cells.

Mitosis – Division of a *diploid* cell resulting in the production of two identical daughter cells.

Notes and references

Mutation – Change in DNA. The term is commonly used for a disease-causing change.

Natural selection – Process through which the environment permits the preferential survival of the better adapted to it.

Nucleic acid – Acidic material originally isolated from cell nuclei and which turned out to be made of *DNA* and *RNA*.

Nucleotide – Basic unit of a chain of *DNA* or *RNA*

Phenotype – Visible trait generated by a *DNA variation*.

Polymorphism – Genetic *variation* occurring with a certain frequency (arbitrarily fixed at 1%) in a population.

Recombination – Exchange of segments between two DNA chains or chromosomes.

Replication – Process through which a DNA chain is duplicated.

RNA – **R**ibo**N**ucleic **A**cid. It is the single-stranded transcript of DNA. It has a cellular function. Its presence within the cell is limited in time, but renewable. It serves as material of heredity for some viruses.

SCD – See Drepanocytosis.

Selective sweep – Transgenerational conservation of a large segment of *DNA* encompassing a variation that confers a survival advantage.

Sequencing – Determination of the succession of units that make up a *DNA*, *RNA*, or protein chain.

SNP – **S**ingle **N**ucleotide **P**olymorphism. Variation of a single unit of a DNA chain.

Somatic - Is said of a trait that is typical of, or a change that affects bodybuilding cell (by opposition to germline or reproductive cell).

Splicing – Process of formation of functional *RNA* involving the removal from the primary transcript of segments corresponding to *introns* and the abutting those corresponding to *exons*.

Transcription– Copying of the information contained in *DNA* in the form of *RNA*.

Translation – Conversion of the information carried by an *RNA* into a protein.

Variation – Any change in a *DNA* chain

Zygote – *Diploid* cell generated by the fusion of two *haploid gametes* which will develop into a full blown organism.

INDEX

Abeba, 70, 72; Mother of humanity, 72

Abebe, 70; Father of humanity, 71

Aborigines, 68; genetic pool, 68

Africa: causes of underdevelopment, 133; the cradle of humanity, 69

African ancestry, 11, 13; African descent, 16, 26, 102

African-American, 11, 12, 81

Africans: education, 135; life priorities, 121; no finger pointing, 133; youth, 137

albinism, 32, 33

allele, 56

Aurelius, Marcus, 88

Bakongo: endogamy permission, 32; twin enfanticide, 39

Baluba, 37; acceptance of congenital abnormalities, 40; cosmogony, 99; culture, 24; endogamy prohibition, 32; terms of emotions, 106;
unusual births, 39; vocabulary of sexuality and reproduction, 33

Barker, David, 61

Bell Curve, 90; intelligence of Africans, 90

Big Bang, 63

Bijimine, Mputu Grégoire, 11

Binet, Alfred, 122

Biologism: the new religion, 85

blood types, 79

brain: seat of the intellect, 88, 92

cells: cytology, 49; definition, 47; gametes, 48; germ cells, 48, 74; division, 48; microscopic evidence, 47; somatic cells, 48, 56, 74; division, 48

CHLCA (*Chimpanzee-Human Last Common Ancestor*), 65

chromosomes, 48; autosomal, 48; homologous, 74; in mitosis, 48; inter-species homologies, 54; microscopy, 49; structure,

48; telomeres, 107; X Chr, 71; X Chr, 71; Y Chr, 71; Y Chr jokes, 71; Y Chr SNPs in sub-Saharan Africans, 71; Y Chr variations, 70

Collins, Francis, 54

congenital anomalies, 40

craniometrics, 89; and intelligence, 89; of women and Africans, 89

Crick, Francis, 50, 51, 94

CRISPR/Cas9 technology, 22, 116; bonanza for eugenicists?, 117; genetically modified babies, 117

Darwin, Charles, 45, 46, 73, 84; and the emergence of species, 46; and the theory of evolution, 45

de Duve, Christian, 130

deoxyribonucleic acids. *See* DNA

Diamond, Jared, 133

DNA: antiparallel strands, 51; as a pharmaceutical, 116; as a vault, 23; as epistles, 125; community of life, 113; DNA sequencing, 12; man's mastery of DNA, 129; molecule of heredity, 50; nuclear DNA, 52; replication, 51, 52; sequencing, 53

DNA double helix, 50, 51

Draper, Wickliffe Preston, 91

drepanocytosis, 37, 108; genetic therapy, 114; symptoms, 109

Earth: Africa is the first human settlement, 73; age, 63, 65; cataclysms, 60; human peregrinations, 66, 78; human peregrinations on, 57; topology, 80; triumph of Man, 65

Ectrodactyly, 39

Edwards, Antony Williams Fairbanks, 79

Egypt: grandeur and religion, 97; kinship with sub-Saharan Africa, 97, 98, 133

endogamy, 32; Vadoma tribe, 39

environment, 25; and ideologies, 121; and intelligence, 122, 123; breeding ground of genetic variations, 60; destruction, 129; gene expression, 73; gene-environment interactions, 60, 65

enzymes: blood, 79; catalysts of life, 47; in DNA replication, 51

epigenetic markings: and gene expression, 73; in germ cells, 61; in zygotes, 61

Index

eugenics: and the theory of evolution, 84; damages and depravations, 87, 88; definition, 83; human ranking, 88; in African culture, 31; instinctive, 84; in mouse, 83; institutional, 84

eukaryotes, 47

Evans, Gavin, 126

evolution: 'benevolence', 110; and the brain, 90; blind or directed?, 130; combinatorial game, 55, 56; *junk DNA* as experimental ground, 54; *Law of parsimony*, 56; ponctuated equilibrius, 60; preservation of pathogenic mutations, 109, 110; presumed favorites, 26

exons, 55

Galton, Francis), 84

gametes, 48; in sexual reproduction, 48

gene: definition, 53

gene-environment interactions, 27, 74, 80, 120

genetic drift, bottleneck, or *founding effect,* 74

genetic engineering: manufacture of biologicals, 114

genetic manipulations, 83

genetic pool, 57

genetics, 8; abused or abusive science, 123; abuses, 26; and race, 78; as a religion, 125; birth of molecular genetics, 51; daughter of heredity, 24; fear of, 12; mythologies, 18, 23, 25; of classifying and ranking, 26; of human nature, 73; perversion, 125; product and producer of culture, 25

genome: a collective of genes, 53; divergence/species emergence, 55; number of genes, 54; sequencing, 54; topology, 80, 81

genotype, 58

Gilbert, Walter, 12, 53

Gobineau, Joseph Arthur de, 85

goiter, a beloved deformity, 38

haplotype, linked variations, 57

He, Jian-kui, 117

heredity, 29; by naming, 35; by reincarnation, 34, 35; definition, 29; of life force, 36; of power and know-how, 35

Hereros and Namas, 87

Herrnstein, Richard J., 90

heterozygous, 57

Hitler, Adolf, 88

homininian, 65

Homo denisova, 66, 69; trace DNA in Asians, 69

Homo habilis, 66

Homo neanderthaliensis, 22, 66; trace DNA in Eurasians, 69, 102

Homo sapiens, 66; agent of evolution, 129; cognitive brain, 93; emergence, 25, 30; pluricentric hypothesis, 69; emergence of, 65; oldest fossils, 69; pre-*Homo sapiens*, 69

homologous recombination, 74

homozygous, 57

hypodescendance, 104

Igbos, 36, 37

Institut de recherches cliniques de Montréal. *See* IRCM

intellect, 92

intelligence, 90; definition, 92

introns, 55

IQ, 88; 'inter-racial' differences, 102; in monozygotic twins, 103; of Africans, 93; of Ameridians and Africans, 15; origin of tests, 122; sterilization criteria, 86; test, a measure of intelligence?, 92

IRCM, 12, 13, 119

Judson, Horace F., 50

junk DNA, 54, 56

Kamenga Mbikay, 7, 16, 17

Kipling, Rudyard, 132

Lewontin, Richard, 79

life kingdoms, 29

Luther King Jr., Martin, 127

Lynn, Richard, 100

Mailloux, Pierre, 14, 15, 16, 17

malaria, 108

Man-chimpanzee as evolutionary relatives, 46, 88

man-chimpanzee, evolutionary relatives, 65

Mansa Moussa, 98

meiosis, 48; homologous recombination, 74

melanin: a pigment, 76; *as a neurotoxin?*, 14; genes, 76, 77; melanocytes, 76; relation to dopamine, 14; screen against UV rays, 76, 77; selective sweeps; around genes, 76; SNPs around genes, 76

Mendel, Gregor, 44, 45, 46; Father of genetics, 45

microbiome, 113

mitochondria: as bacterial colony in eukaryotic cells,

Index

52; cellular furnaces, 52; single female ancestor, 72

mitochondrial DNA. *See* mtDNA

mitosis, 48; microscopy, 49

molecular clock, 57

Monod, Jacques, 130

Monomotapa, 97, 98

mtDNA, 72; exclusively maternal heritage, 72; genes, 72; SNPs, 72

Murray, Charles A., 90

mutation, 39; an asset or a liability, 108; as an experiment of Nature, 108; definition, 58; from DNA replication, 51; pathogenic variation, 58

natural selection: "original sin", 131; as a means to evolution, 45; as a source of all supremacies, 85; as an ecological equilibrium among species, 67; circumventing its traps, 136; cleansing of pathogenic variations, 58; gene meanderings and sorting, 17; its signatures, 58, 73; of skin pigmentation, 76; preservation of pathogenic mutations, 109; purifying effects, 84; reinterpreted, 107, 129; selective pressures, 57; selective sweeps, 74, 75, 77; survival tests, 64; the three life imperatives, 67

nucleic acids, 49

nucleotides: structure, 49

Nzinga Mbandi, 97

ogbanje, 36, 37

OoA. *See Out of Africa*

Ota Benga, 87

Out of Africa: monocentric hypothesis, 69

pea plants, 44

phenotype, 58; due to geographical or cultural isolation, 74; morbid, 58

Pinker, Steven, 120, 127

polymorphisms, 56

prokaryotes, 47

race: absurd classifications, 80; an induring illusion, 119; arbitrary choice or imposition, 80; race is an illusion, 78, 79, 118; social value, 79; what is it?, 78

regression of averages, 104

reincarnation, 37

ribonucleic acids. *See* RNA

r-K strategy, 105

RNA, 50, 116

Rushton, J. Philippe, 89; on 'inter-racial' IQ differences:,

100; on Africa connectivity, 99; on Africa's backwardness, 96; on African reproductive drive, 105; on brain size, 91; on criminality, 99; on IQ heritability, 102, 104; on the Bell Curve, 90; source of research funding, 91

Saini, Angela, 122

Samory Touré, 98

Sandra Scarr's Minnesota Project, 103

Sanger, Fred, 53

Sarah Baartman, the Hottentot Venus, 87

SCA. *See* sickle cell anemia

SCD. *See* sickle cell disease

Schopenhauer, Arthur, 128

science: abused and abusive science, 125; and race, 78; and the truths, 26; antiracism, 79; contemporary mythology, 23; definition, 26, 43; *Law of simplicity*, 70; medical anthropology, 112; of reproduction in African culture, 34; perversion by ideology, 96; proto-science, 43; sacralization, 26; sources of controversies, 80; the exception invalidates the rule, 87

sexual reproduction, advantages, 49

Shaka Zulu, 97

Sheikh Anta Diop, 98

Shockley, William, 86

sickle cell anemia. *See* drepanocytosis; as a shame, a stigma, a curse, 14

sickle cell disease, 37, 108; *bokono ya kibeka*, 38

SNPs: beneficial, 75; geographic distribution, 68; neutral, 75; single nucleotide polymorphism, 56; sub-Saharan Africa, 68; tracking migrations, 67

species, 117; as cathedrals of genomes, 58; definition, 29; designed improvement, 84, 85; emergence, 46; improvement, 86; migrations, 66; selective pressures, 67; types, 30; animals, 30; humans, 30, 31

splicing, 56; alternate splicing, 56

Stephen Jay Gould, 60, 123

sub-Saharan Africa, 133; fear of genetics, 23; human types and subtypes, 31; migrants, 132; President Trump's opinion, 145; rades by Arabs and Europeans, 96; the S mutation and malaria, 109; underdeveloped, 131

Index

Sun, age, 63, 65

Sundiata Keita, 97, 98

synteny, 54

Theilard de Chardin, Pierre, 130

theory of evolution: basis of research on IQ, 95; mythologies, 25

transcription, 52, 55; transcriptional factors, 59

translation, 52; post-translational modifications, 59

Tree of Life, 25, 46, 64; *phylogeny*, 65

Universe, age, 63, 65

Vadoma tribe, Lobster People, 38, 110

variations, 56; caloric parsimony, 111; caloric prodigality, 111; causes of, 57; chromosomal rearrangements, 56; hereditary, 57; insertions/deletions, 56; neutral, 57; paternal and maternal combined, 74; pathogenic, 81; renal salt retention and high blood pressure, 111; substrates of Natural Selection, 57; wisdom SNPs, 130

Venter, Craig, 54

vitamin D: manufactured by skin, 76; production stimulated by solar rays, 77

Watson, James, 50, 51, 94; on the IQ of Africans, 94

Wilson, Edward O., 105

zygote, 61; the formation of, 48

www.ingramcontent.com/pod-product-compliance
Lightning Source LLC
Chambersburg PA
CBHW070641220526

45466CB00001B/245